Electric Power Generation

A Nontechnical Guide

Electric Power Generation

A Nontechnical Guide

by Dave Barnett
& Kirk Bjornsgaard

PennWell Books
PennWell Publishing Company
Tulsa, Oklahoma

Cover design by Shanon Garvin and John Potter
Book layout and design by Brian Firth

PennWell
1421 S. Sheridan Road/P.O. Box 1260
Tulsa, Oklahoma 74101

Library of Congress Cataloging-in-Publication Data
Barnett, Dave and Bjornsgaard, Kirk
Electric Power Generation: A Nontechnical Guide\Dave Barnett and Kirk Bjornsgaard
p. cm.
Includes index
ISBN 0-87814-753-5

Printed in the United States of America.

04 03 02 01 00 5 4 3 2 1

Dedication

For John Turnicky, my first mentor; Mr. Schmertz, the science teacher who started my technical writing career by making the point: for full credit on a 10-point essay question, put in 10 facts; Rod Luck and Len Smith; Frau Voltz, *wer mit "minus Punkte," SpaB. usw, machte etwas der Forschung dieses Buch moglich*; Robert A. Heinlein; Jean Shepherd; Mrs. Ann O. Hess, whose "label-factor method" got me through both her class and, with my HP 33E, college chemistry; Joe Wiley; Ray Kiesel, and Alvin See, who shared the secrets of control system design; my parents, David J. Barnett, Jr. and Arlene Mann Barnett; my sons Zach and Alex Barnett; Tristan Chambers; Kirk Bjornsgaard, my best friend from high school and now my co-author and editor; Marlene Leach, my love; and my Higher Power, whom I choose to call God.

—Dave Barnett

For Helen Ross, John Harrison, and Paula Munson, who exemplify what good teachers can be; for Rod and Len; for my dad, Frank Bjornsgaard, and his unswerving encouragement, and for Noma Krasney, Ph.D., for her work in reviewing the manuscript and her wholehearted support.

—Kirk Bjornsgaard

Table of Contents

Figures List

Tables List

Acronyms List

A	Amperes
ABWR	advanced boiling water reactor
ac	alternating current
ACSR	aluminum clad steel reinforced
APWR	advanced pressurized water reactor
B&W	Babcock & Wilcox
BIL	basic impulse insulation level
Btu	British thermal unit
BWR	boiling water reactor
CAA	Clean Air Act
CAAA	Clean Air Act Amendments
CANDU	Canada deuterium uranium reactor
CEA	Commissariat ê Energie Atomique
CETI	Clean Energy Technologies, Inc.
cgs	centimeter-gram-second
CO_2	carbon dioxide
CPR	cardiopulmonary resuscitation
CWA	Clean Water Act
D_2O	"heavy water"
DBS	direct broadcast satellite
dc	direct current
DOE	Department of Energy
DSS	daily start-stop
EAP	employee assistance programs
EHV	extra high voltage
EIA	Energy Information Agency
EMF	electromagnetic force
EMS	emergency medical system
EMT	emergency medical technician
EPA	Environmental Protection Agency
EPRI	Electric Power Research Institute
ETBE	ethyl tertiary butyl ether
F	Fahrenheit

FERC	Federal Energy Regulatory Commission
FMEA	failure modes and effects analysis
GCA	ground controlled approach
GE	General Electric
GW	gigawatt
H_2	hydrogen gas
Hz	Hertz
IOU	investor-owned utilities
J	Joule
kg	kilogram
kHz	kilohertz
km	kilometer
kPa	kilopascal
kV	kilovolt
kVA	kilovolt-ampere
kVAR	kilovar
kW	kilowatt
kWh	kilowatt-hour
LAN	local area network
LENR	low-energy nuclear reaction
LEO	low earth orbit
LTA	logic tree analysis
LWR	light water reactor
MAD	mutually assured destruction
MASER	microwave amplification by simulated emission of radiation
Mcf	thousands of cubic feet
MGN	multi-grounded neutral
MHz	megahertz
MKS	meter-kilogram-second
MkWh	million kilowatt hour
MNO_2	manganese dioxide
MOX	mixed oxide fuels
mpg	miles per gallon
MSW	municipal solid waste
MTBE	methyl tertiary butyl ether
MVA	mega Volt Amperes

MVAR	mega Volt Amperes reactive
MW	megawatt
N_2	nitrogen
NAAQS	National Ambient Air Quality Standards
NASA	National Aeronautics and Space Administration
NEC	National Electrical Code
NIMBY	"not in my backyard"
NO_x	nitrogen oxides
NPDES	National Pollutant Discharge Elimination System
NRC	Nuclear Regulatory Commission
O&M	operating and maintenance
O_2	oxygen
OPEC	Organization of Petroleum Exporting Countries
PC	personal computer
PCS	personal communications services
PM	predictive maintenance
psi	pounds per square inch
PV	photovoltaic
PVC	polyvinyl chloride
PWR	pressurized water reactors
RBMK	Russian acronym for high power channel reactor
RCM	reliability centered maintenance
rpm	revolutions per minute
s	seconds
S	Siemens
SF_6	sulfur hexafluoride
SI	System International d' Unites (SI metric system)
SIP	state implementations plans
SO_2	sulfur dioxide
SO_x	sulfur oxides
SPSS	solar power satellites system
Sv	Sievert (equivalent radiation dose)
T	Tesla
TMI	Three Mile Island
UHV	ultra-high voltage

UPI	United Press International
V	Volt
vac	vacuum
W	Watts
WMEC	Western Massachusetts Electric Company

Introduction: A Change in the Air

Notice anything different lately about your electric power? Your lights still burn brightly. The refrigerator still keeps food cold. This season your television hypnotized you with sparkling new images and content not much different from last year. Your alarm clock still wakes you before you have had enough sleep.

If you checked the power in a receptacle with a digital multimeter you would find it is precisely the same old 120 Volts, 60 cycles per second (Hertz) alternating current. The only clue might be your electric bill. Is it a little less than last year? Maybe not.

The change of which we speak is comparable to the changes in the telephone industry that led to the breaking up of the Bell system. It is competition. Deregulation. A similar, very important change in technology drives the current competition and deregulation of the electric power industry. In the telecommunications industry, the driving technology was the transistor, followed by integrated circuits and microprocessors. In the electric power industry, natural gas-fired turbine generators—essentially, jet engines directly connected to electric generators—are changing the economy of scale for electric power generation. These gas-fired "prime movers" eliminate the middleman of the old steam boilers to turn the generator on the same shaft, increasing performance and efficiency. As will be seen, applying the turbine exhaust to a steam turbine generator still leaves plenty of steam for manufacturing processes or heating.

The "new" technology of gas turbine generators (that has been around since the end of World War II!) will allow reasonable environmentally friendly plants to be sited where the heat of "waste" steam can be put to good use. Burning natural gas in the air produces water vapor and carbon dioxide (CO_2)—the latter a "greenhouse gas"—as combustion products. Side reactions create nitrogen oxides (NO_x) and some sulfur oxides (SO_x). Once big, old fossil-fuel plants and nukes are sold to recover their owners' "stranded costs," they will economically handle the base load electrical generating capacity year in and year out, just as they always have.

The early promise of cheap long-distance calls, made when Ma Bell was broken up, was offset by the sheer hassle of having to deal with multiple telephone equipment and service providers. However, Humpty Dumpty could not be put back together again. For more than a generation, telecommunications deregulation has been a real economic advantage to perceptive business and residential customers. Only in the second half of the 1990s did the average person feel it—in the impact of cellular phones, the Internet and World Wide Web access, pagers, and so forth. Technological advances made these financial, life-, and work-style benefits possible after deregulating and dismantling what was once thought of as a "natural monopoly."

The economy of scale of the large fossil-fuel, nuclear-fired, hydroelectric generating plant has been challenged by the moderate-sized gas turbine generators that can be built on-site for a customer requiring both electric energy and steam for manufacturing processes or for heating. Will it take 25 years for the break-up and re-regulation of the electric utilities into generating, transmission, and distribution companies to be perceived as a net benefit by managers and consumers alike?

Two things are certain: there is nothing so constant as change and the more things change, the more they stay the same.

This, then, is our adventure: how changes in technology have changed—are continuing to change—the generation of electric power.

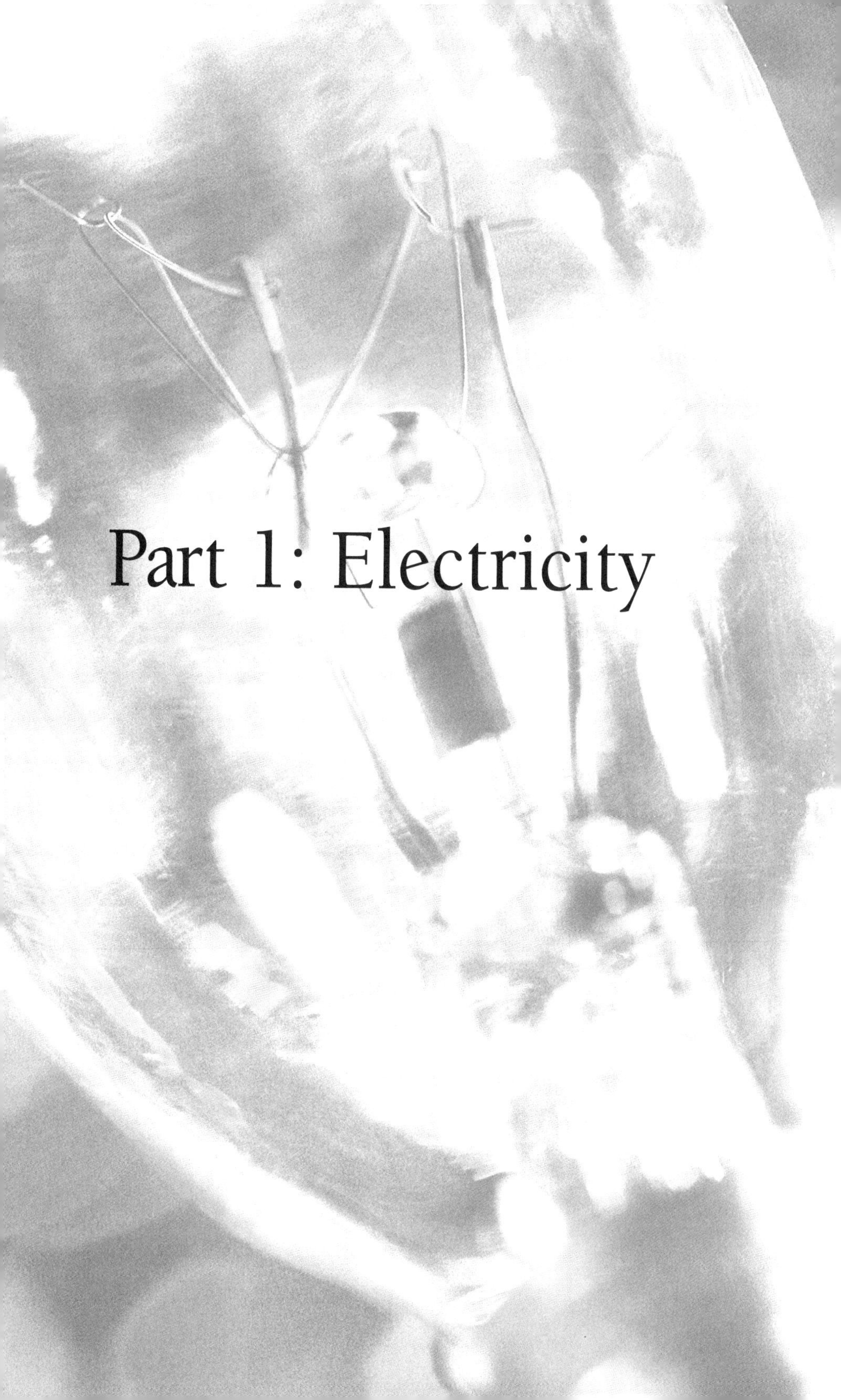

Part 1: Electricity

Chapter 1
From Frogs Legs to Microwaves

"Everything I told you is a lie."

— Alan A. Jones, Ph.D.
Chair, Chemistry Department, Clark University

The history of science does not change even as engineers and scientists add historic breakthroughs at an ever-increasing pace. Maybe that increasing pace is the reason why high school and even college science courses mainly teach the history of science, rather than actual science. To a new generation born in the third millennium, the incredible technological advances of the Twentieth Century may blur together with the series of scientific discoveries that occurred earlier with the exploration and colonization of planet Earth by European countries. Change is not politically correct.

Atoms are not little plastic beads that snap together into molecules like a child's toy. Atoms are often described as mainly empty space, the hydrogen atom nucleus compared to a baseball at home plate, with a lone tiny electron a third of a mile away. With this image in mind, it is easy to believe the comic book hero Flash could vibrate his molecules past those in a wall, or even take a shortcut through the whole Earth!

In reality, atoms are not made of white marbles for protons, gray marbles for neutrons (white plus black equals gray), and black dots for electrons. Whether you are in the infield, outfield, parking lot or stands, you had better look out! Electrons move in complex orbits, at nearly the speed of light—seemingly everywhere at once.

However, just as classic Newtonian physics is good enough for most National Aeronautics and Space Administration (NASA) orbital calculations, the Neils Bohr model of the atom is good enough for our discussion. Just keep in mind while the history is OK, contemporary scientists have proven physics is a lot more complicated than our bead model and everything (theoretical) we are about to tell you is a lie!

Those frog legs

This story begins in 1752 in colonial America, where (fortunately) Benjamin Franklin survived his Philadelphia experiments with lightning. Franklin's work was not the last in electrostatic investigations that began in antiquity with rubbing fur across amber rods, but it marked a departure that would lead to practical applications and tremendous advances understanding electricity and energy.

In 1780 in his laboratory in Bologna, Italian physician Luigi Galvani performed a series of experiments to determine the effect of static electricity on animal tissue. In one experiment, Galvani and his assistant observed a dead frog's leg twitching when stimulated by a nearby electrical machine operated by another experimenter. In a fateful test, Galvani intended to stimulate frog legs with the charge from a lightning rod invented by Franklin. If he had hung the frog legs from the cast iron railing with hooks made of iron wire rather than brass or copper, he might have developed the electric Hot Dogger nearly two centuries before Presto did and never have made the discovery that galvanized the scientific world. Even without static electricity from the lightning rod, whenever wind caused the legs to contact the bars supporting the railing, the muscles were stimulated to contract. Dr. Galvani's "animal electricity" experiments led him to conclude animal tissues contained an "electric fluid."

From the bifocal glasses we wear to the wood stoves that warm us during cold winters, many of Franklin's inventions are manufactured today with only improvements in materials and methods. However, as important as Franklin's experiments with lightning were, they only set the stage for Galvani and the electrical researchers who followed.

Galvani's experiments moved knowledge of electricity beyond the Greeks' speculation about atoms and his contemporaries' experiments with static electricity. With the development of the electric circuit, science advanced considerably and paved the way for practical applications. Recently butchered frog muscles contracted to indicate the electric current passing through them. The frog leg hanging by a hook made of a metal dissimilar to the iron railing and coming into contact with the railing, formed a battery with its terminals shorted together—the first short circuit.

Italian physicist Allesandro Volta showed that Galvani's frog legs were both an indicator of electric current flow and provided the electrolyte—a vital part of a battery. The neuromuscular tissues in the legs provided the visible response, while the salts dissolved in the blood and cells of the frog leg were the electrolyte. While Galvani's focus was on the legs of the animals he used, the brass or copper hooks and the railing were the electrodes. He had created the first artificial electric circuit. Long before Mary Shelley wrote the novel *Frankenstein or The Modern Prometheus* in 1816, Volta had demonstrated that living or previously living tissue was not required for electrical experiments.

Volta developed both "dry" and wet cell batteries. Dry cell batteries are not really dry; they use a small amount of paste-like electrolyte or liquid absorbed into fiber separators. Electrodes of two distinctly different materials, (such as carbon and tin or copper and zinc) are separated by paper soaked in a salt solution to form the dry cell battery known as a voltaic pile. For anyone who has taken apart the traditional "cat" battery, the anode or positive electrode is a mixture of carbon granules and manganese dioxide (MNO_2) that surrounds the carbon rod, which is a compatible conducting material, rather than the electrode itself.

Volta's battery of wet cells was called a "crown of cups." The circular arrangement of the cells in individual glass jars merely allowed the battery terminals to be close to each other.

Strictly speaking, a battery consists of several cells, packaged together and normally connected in series with negative (-) terminals to positive (+) terminals to increase the voltage output. However, single cells have been so commonly called "batteries" that the usage is now acceptable. The six 2.2 Volt (V) lead-acid cells in the rectangular plastic case under the hood of your car are properly called a battery; however, the actual fully charged voltage is about 13.2 V—so it is not really a "12 V" battery.

In addition to wet and dry cells, batteries are also classified as primary and secondary types. Primary cells, including camera and flashlight batteries are used once and thrown away. Because they are not recycled, primary cells for consumer, commercial, and industrial use must not contain significant amounts of cadmium, lead, mercury or other heavy metals or toxic materials under current environmental policy. Secondary types are rechargeable. These include the lead-acid batteries used in vehicles and boats, the nickel-cadmium and nickel-metal hydride batteries used in camcorders, pagers, portable two-way radios, and radio-controlled cars.

Environmental regulations and sound economics coincide—for the most part—with regard to rechargeable batteries. The "heavy" metals used in rechargeable battery chemistry—lead, cadmium, etc.—are also cheaper to recycle than to mine and smelt, making the battery recycling box at your local Radio Shack more than just a good idea. Recycling the components of old batteries makes new batteries less expensive and protects the environment.

Contrary to conventional wisdom, both old-fashioned, guaranteed-to-leak-eventually carbon-zinc and alkaline "primary" cells are rechargeable—with serious limitations. The metal can surrounding carbon-zinc cells is also the negative electrode and is consumed in the process of converting potential chemical energy to kinetic electric energy. (Better quality batteries also have an impregnated kraft paper or plastic seal and outer steel jacket to delay leakage when the electrode is used completely.) In alkaline cells, making electricity consumes the central electrode. Again, contrary to conventional wisdom, the electrolyte is not an acid. In carbon-zinc cells, the electrolyte is a combination of two salt solutions, zinc chloride and ammonium chloride. In "alkaline" batteries, it is potassium hydroxide, a strong base. When the zinc electrode disappears in a carbon-zinc cell, the electrolyte released will cause battery contacts and other metal parts to rust or corrode as electrodes. Recharging either type will not restore the consumed electrode to its original physical dimensions, limiting both peak current and total energy. Therefore, by the time they need to be recharged, they cannot be!

Recharging these batteries also releases hydrogen gas that may cause them to rupture. The gas is flammable. The current generation of low-cost zinc chloride batteries have a longer shelf life than the ones you used to get free with your Radio Shack battery card, but like other primary batteries—including lithium, silver oxide and zinc-air—they really are not rechargeable. Trying to recharge any of these types may cause burns or rupturing of the cases and they will not provide reliable power. If you use a lot of batteries, rechargeable batteries and a charger pay for themselves after about five charging cycles.

Rayovac ran afoul of the letter of environmental law when it introduced "renewable" alkaline batteries. Federal regulations required that all rechargeable batteries be recycled. Even though the Rayovac batteries did not contain mercury or cadmium, etc., were safe in the domestic waste stream, and were designed to do their part to reduce that waste stream, they did not meet federal requirements.

It is not where you draw the line, but how you draw the line. Apparently the difficulties have been ironed out and the product is not only successful but recommended for use by government agencies and available to them through the General Services Administration!

Beyond batteries

While theoretical physics is well beyond the scope of this primer, within their limitations analogies serve us well for how electricity behaves. The analogy to water works on several levels. Atoms and subatomic entities, such as electrons, sometimes behave like the marbles we have been taught to imagine. Just as often (going back to the water analogy) they behave like waves.

All analogies have significant limitations. Water must ultimately travel in a complete cycle—from sky or stream to a reservoir, or from a well through pipes to point of use and then through septic or sewer systems into the ground or ocean—to eventually replenish the source in order to perform useful work. Electric currents must *immediately* pass through a complete circuit from one terminal of the battery or generator, through a conductor to the load, and then back through an associated conductor to

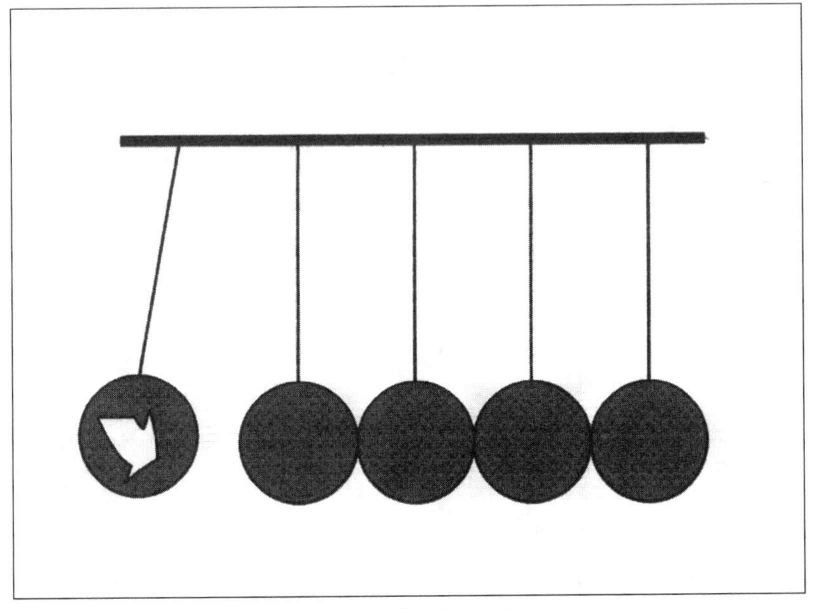

Figure 1-1a: How Energy is Transmitted within a Conductor

the other terminal of the electric energy source. Static electric discharges, such as lightning strikes or touching a wall switch plate after scuffing across a carpet in dry winter weather, can occur without a complete electric circuit. However, only electric power traveling in a complete circuit can do useful work.

Figures 1-1a, b depict how energy is transmitted within a conductor. Like ball bearings hanging from strings, individual electrons do not have to move very far to conduct energy. Because of the arrangement of electrons in conductors such as copper, gold, silver—even aluminum, mercury and iron—the electrons easily transfer electrical energy from one atom to another.

Think of the molecular structure of insulating materials as being arranged so electrons cannot carom off each other and conduct an electric current. Nonmetallic solids that do not dissolve in or absorb water (such as ceramics, quartz, plastic, rubber, or wood) tend to make excellent insulators. Their electrons are locked in a lattice structure and are unable to conduct a current. Organic compounds, whether gas, liquid or solid—

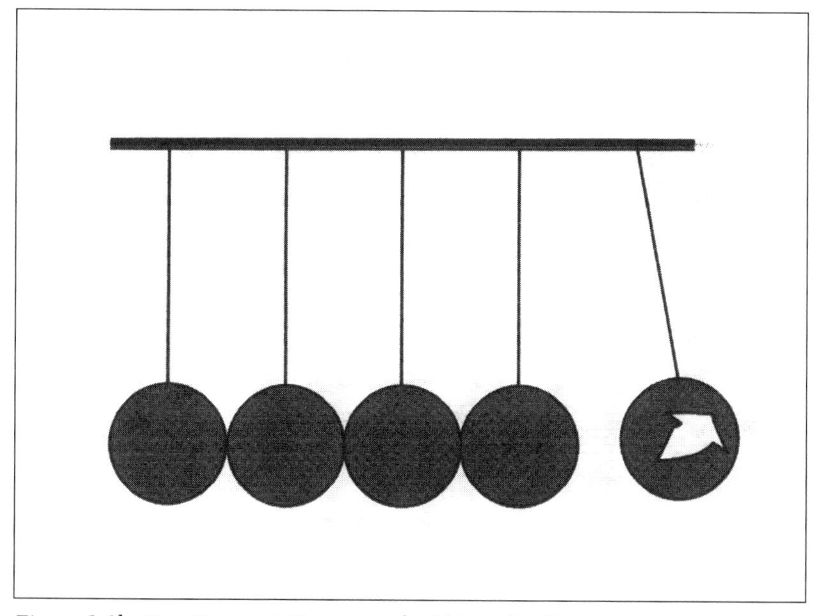

Figure 1-1b: How Energy is Transmitted within a Conductor

again, if they do not dissolve in water—generally do not conduct electricity. Gases usually make good insulators. "Noble" gases such as helium, neon, argon, and krypton have their outer electron shells filled and, as in the case of neon signs, conduct a current only when exposed to high voltages. More mundane gases like CO_2, nitrogen (N_2), and oxygen (O_2) exist as molecules with their outermost electrons locked in stable arrangements, normally unable to conduct electricity. The air we breathe is an adequate insulator for many applications. Hydrogen (H_2) gas conducts heat so well, has such low viscosity, and is so stable in the presence of high voltages, it is used as a coolant in most larger generators. Sulfur hexafluoride (SF_6) gas insulates many underground cables.

Diodes, transistors and integrated circuits use semiconductors—a special case of insulators. Pure germanium and silicon are insulators. However, when these materials are selectively "doped" with small amounts of gaseous aluminum, arsenic, gallium, indium, etc., at high temperatures, they form regions with an excess of electrons ("n" for negative type) or an electron deficit ("p" for positive type) that can be made to conduct electric currents. An "n"-type region sharing a junction with a "p"-type forms electricity's one-way valve, called a diode or rectifier. A small region of "n-" or "p"-type material separating two larger regions of the opposite type constitutes a sensitive electrical flow control valve or on-off switch known as the transistor. Many diodes and transistors on larger "chips" form integrated circuit amplifiers, digital logic circuits, memory, or microprocessors. Certain ceramic materials have been found to act as semiconductors at high temperatures, leading to the development of electronics that operate in furnaces or even in spacecraft traveling to the surface of Venus.

Electrifying developments

With a new source of electrical energy that did not depend on bad weather, frog legs, or kites, experimenters and inventors quickly developed devices to use that energy. From the very beginning of this era, there were two categories of electric equipment with two types in each:

- power vs. signal
- analog vs. digital

Operating the electric motors and incandescent lights that would soon be developed required brute force power, while telegraphs and telephones required much smaller signal currents. Digital calculators, computers and music CDs did not become widespread until the last quarter of the Twentieth Century; however, the earliest electric communications system, the telegraph, was digital.

The Industrial Revolution was already benefiting from compressed air-, steam- and water-powered machinery. Modern technology was advancing agriculture, food processing, business, education, entertainment, manufacturing, medicine, printing, textiles, transportation, and warfare. Lives could be improved, or could be ended more efficiently and violently. Electricity was just one more development in a world of chemical and mechanical engineering.

People today are highly sophisticated in the use of technology, but are we "smarter" than our predecessors? Greek and Roman philosophers articulated concepts about physical and spiritual aspects of the world.

Archimedes was the first scientific detective, determining that a crown was a less dense alloy of gold and silver by the fact it displaced more water than an equal mass of pure gold. Sophisticated thought was advanced with the commandment to "love one another" and the radical notion to "turn the other cheek" when attacked by our enemies. During the Renaissance (early 1300s to late 1500s) and in William Shakespeare's plays we see the beginning of modern points of view. Gottfried Wilhelm von Leibniz (1646-1716) and Sir Isaac Newton (1643-1727) had developed calculus by the third quarter of the Seventeenth Century. It is difficult to imagine just how much stuff had to be invented. Once basic principles were arduously worked out and materials became available, advancements snowballed.

In 1821, Englishman Michael Faraday invented the electric motor that ran on batteries. William Sturgeon made the first electromagnet in 1825. Today, any kid with some insulated wire, a nail, and a battery can do that. In 1830, American Joseph Henry developed the transformer. By 1831, Faraday and Henry, working separately, had established the principle of electromagnetic induction, which led to the electric generator. While communication devices such as the telegraph, and later, the telephone, could—and, until the middle of the Twentieth Century many did—run on batteries, the generator made practical the incredible array of higher powered inventions to follow.

The following inventions were not just better mousetraps. Each one spawned an entire industry.

Samuel F. B. Morse and Alfred Vail's 1844 recording telegraph receiver and alphanumeric ("Morse") code was not the first commercially viable telegraph system. In England, in 1837, William F. Cooke and Charles Wheatstone implemented the first railway deflecting needle telegraph. The Wheatstone bridge uses a series-parallel network of resistors or sensors to isolate microvolt signals from interference-laden environments. The Wheatstone bridge provides the analog gateway to modern digital sensing systems.

Morse designed his receiver to record automatically the dots and dashes used in his code on a paper tape. This first digital marvel used both electric hardware and "wetware" (software resident in the minds of the operators). Ironically, his receiver made such a good sounder, operators could easily distinguish "dits" (dots) from "dahs" (dashes) and wrote messages in plain English, eliminating the benefit of the tape record. Morse's other innovation, also later abandoned for communication purposes, was to save half the wire required to complete the circuit by using the earth as the return conductor. Ground rods were first used in Franklin's lightning rod system. Copper or copper-flashed iron rods at least eight feet long driven into good soil provide a low-resistance connection to the Earth. Contrary to conventional wisdom, electricity does not just follow the path of least resistance, it follows *every* path. The earth is a virtually infinite parallel network with its conductivity limited only by the soil surrounding the ground rods and the length and depth of the ground rod itself. Therefore, if you can make a good connection to the ground, you have a free "wire" as long as you want that will carry as much current as you want.

As traffic increased, additional wires were strung on crossbars of poles installed along railroad rights-of-way, roads, and city streets. Eventually, as multiple wires were bundled into cables—especially after the development of the telephone—"crosstalk" became a problem and the single wire circuit using the earth ground as a return gave way to twisted-pair wiring. Twisted-pair wiring is inherently self-shielding and is used today in computer local area networks (LANs) as well as standard telephone wiring. Cabled wiring can be of lighter gauge and therefore cheaper because mechanical strength and support can be separate from the conductors.

The forerunner to today's Internet was the telegraph network. Telegraph systems were party lines with many operators at stations along railroad lines and at telegraph offices. In addition to revenue-generating traffic, telegraphers could transmit information from their employers or just exchange humorous messages. While requiring sophisticated skills to use and not available to the general public, this was a truly interactive network.

In 1876 Alexander Graham Bell—and others—invented the telephone, but Bell got to the patent office first. The result of his attempt to create a telegraph capable of transmitting multiple messages simultaneously, Bell's telephone was an analog signal system. His original design was similar to the sound-powered phones still used in mines, submarines, and other critical applications. Long before electron tubes or solid state electronics, Thomas Edison developed the carbon granule transmitter, or microphone. The transmitter acted as a variable resistor responding to changes in sound pressure to convert voices to electric signals. The receiver was a thin iron or steel diaphragm supported around its edge with a small air gap between it and a permanent magnet wound with a coil of fine insulated wire. Because the transmitter needed a small amount of current passing through it whenever the phone was in use, and the receiver magnet would be weakened by the talk battery voltage, the two were isolated by a transformer network called a hybrid. The hybrid also allowed just the right amount of the caller's own voice to be heard in the receiver ("sidetone") so people would speak at the correct volume to be heard clearly at the other end. As the telephone system improved, the specifications for the hybrids were occasionally changed. While an electronic home phone of today—set to "pulse"—would work on an early dial system, the sidetone level at your end would be too high and the other person would constantly ask you to speak up.

Ironically, while Bell invented the telephone, his famous assistant, Mr. Watson, invented the bell. Called a "magnetically biased ringer," Watson's brass bell was used in dial and touchtone phones until electronic ringers in retail telephones became popular after the deregulation of the telephone industry. The magnetically biased ringer consisted of an electromagnet wound around a permanent magnet. It was wired into the network with the correct electrical and magnetic polarities to avoid "tapping" when another phone on the line was picked up or dialed.

There were just so many things crying out to be invented. The original Strowger switchgear dial telephone system was invented by an undertaker whose competitor's wife was the town's telephone operator. Fearing his business was being diverted, Strowger leveled the playing field with his more convenient, democratic, and *private* automatic central office.

Thomas Edison, Thomas Edison, Thomas Edison

In 1879, Edison made a carbonized cotton thread filament glow in an evacuated glass bulb for 40 hours. By 1882, Edison was selling electricity generated by steam-driven dynamos from the Pearl Street Power Station in New York City. Edison's insistence on using 110/220 V direct current (dc) dynamos, rather than more efficient higher voltage alternating current (ac) units would later drive him out of the power generating business (Fig. 1-2).

Fig. 1-2: Thomas Edison (Courtesy of EEI)

While experimenting with incandescent lights, in 1883, Edison discovered thermionic emission, also called the Edison effect. He included an additional electrode inside the lamp in the form of a small metal plate facing the filament. He found if the negative terminal of a second battery were connected to one of the filament terminals and the positive terminal to the plate, the glowing filament would emit electrons and a current would flow. Reversing the battery voltage (or bias) did not allow current flow. This discovery would lead to the practical diode tubes useful in low power analog signal electronics for detecting radio waves and high power rectifier versions.

In 1896, Guglielmo Marconi patented his radio telegraph and in 1901 dramatically proved it worked in a transatlantic test from Cornwall, England to St. John's, Newfoundland. Marconi's early transmitters and receivers used brute force power on the sending end and totally passive receivers. Electronics, as we understand the term today, were not involved. Huge generators and transformers created sparks controlled by a telegraph key across a tuned circuit in the transmitter. In the receiver, coils and capacitors tuned to the dominant frequency of the transmitter created a much smaller spark.

If the diode tube invented in 1905 by Sir J. Ambrose Fleming was a check valve, Lee DeForest's 1907 triode was a faucet. Both tubes trace their ancestry to the Edison effect. Described functionally, electron tubes are more properly called valves. (On schematics, their reference designation is "V.") In analog mode, valves amplify signals. In digital or switch mode, they can replace relays and operate thousands or millions of times faster. Both analog and early digital computers used tubes. The triode allowed a small voltage on a grid electrode to control a much larger amount of current that would pass between the cathode heated by the filament and the plate of the tube. These devices signaled the beginning of real electronics. Power diodes could rectify ac power into the dc required by early motors; other electronic devices and signal diodes could detect faint radio signals. Triodes, and later tetrodes (4 electrodes), and even pentodes (5 electrodes) amplified those signals into high-fidelity waveforms reproducible by speakers. Even in the days of digital television, the final amplifiers of UHF television transmitters use tubes because they still have significant efficiency, linearity-of-signal, and power handling advantages over transistors.

In 1906, Reginald Aubrey Fessenden transmitted the first words by radio from the coast of Massachusetts to ships in the Atlantic Ocean. In 1920, radio station KDKA began regular broadcasting from Pittsburgh, Pennsylvania. (It continues to broadcast today!)

The beginnings of television were only three years behind the sign-on of KDKA. Vladimir Kosma Zworykin patented the iconoscope TV camera tube in 1923, and submitted another application for a patent for a color tube in 1925. Economics, politics, and wartime logistics held back television as much as technical limitations.

Scientists make it known, engineers make it work

As this century marched toward World War I, formally trained engineers began to reinforce the progress started by academics, curious amateurs, inventors, and tinkerers. Western Massachusetts native Vannevar Bush, who earned his Ph.D. and served as a professor and later dean of engineering at the Massachusetts Institute of Technology, invented the differential analyzer, a predecessor of the analog computer.

In 1948, Bell Telephone Laboratories' scientists John Bardeen, Walter Brattain, and William Shockley invented the transistor. Like the electron tube, the transistor could amplify signals by varying the bias current on the base electrode, much as high pressure water flows may be controlled by turning the handle on a faucet, or act as an electronic switch. In 1956, they received the Nobel prize in physics.

John Presper Eckert and John W. Mauchly, later in association with John von Neumann, designed ENIAC, EDVAC, and UNIVAC, the first digital computers, built between 1946 and 1958. Their machines used upwards of 18,000 electron tubes and were not available in notebook models. ENIAC had a fixed program for calculating ballistic trajectories. EDVAC's programming could be readily changed as needed. UNIVAC was the first commercially available general-purpose computer. Herman Hollerith's (1860-1929) rectangular punch cards, first invented for the 1890 U.S. census, were used to program computers through the 1970s. Hollerith's tabulating machine company grew into IBM.

Subsequent generations of computers used transistors, integrated circuits, and ever faster and more powerful microprocessors. They evolved from mainframe computers to personal computers (PCs), which then became networked to servers and connected to the World Wide Web. Moore's Law stipulates computing power doubles every 18 months—a 10 billion fold increase since the vacuum (VAC) tube computers of 1950 and an increase of a trillion since the gear-driven fire-control computers of World War I. As the cost of computing power decreases, futurists project microprocessors to be everywhere and in everything from clothing to paper.

Electrons in the large numbers used in today's electric circuits remain as predictable as the outcome of an election with 1% of the vote counted.

As integrated circuits are miniaturized to the quantum level, electrons become as unpredictable as any one voter. For future computer designers, the challenge is to achieve the accuracy, predictability, and reliability of today's microprocessors using a handful of electrons in "nanoprocessors."

Change in the air

Two sides of a coin must be kept in mind as technology evolves and mutates for "there is nothing so constant as change," and "the more things change, the more they stay the same."

Traditionally, signals that required very little bandwidth, such as telephone and teletype, were hardwired. Television signals, which in the U.S. require 6 megahertz (MHz) of bandwidth for each channel (more in Europe and most countries of the world), were broadcast over the air. The FM broadcast band accommodates a hundred radio channel assignments (though not all can be used in any one region, requiring careful "band plans"). The 88-108 MHz FM band, located above VHF channel 6, could only accommodate three TV channels. The entire AM broadcast band, from 535 to 1605 kilohertz (kHz), provides only about a sixth of the bandwidth required for one TV channel. The conventional copper wire twisted-pair telephone line only provides a low fidelity 4 kHz frequency band.

Since cable TV systems were first built in the 1950s and people began buying the first cordless telephones in the '60s, a revolution in communications has been underway. Increasingly, broadband TV signals are being hardwired or delivered by direct broadcast satellite (DBS). Narrowband telephone signals are now more often carried on low power radio waves to cellular and cordless phones.

Clearly, with analog cellular telephones, digital personal communications services (PCS), pagers, and wireless LANs proliferating, it is a waste of finite bandwidth to broadcast television signals over the air when cable TV and DBS systems deliver higher quality signals without consuming scarce bandwidth better used by other services. Satellites not only can reuse the same set of frequencies from "bird" to "bird," with horizontal and vertical polarization techniques, but each satellite uses the *same* channels twice.

Once again, the more things change, the more they stay the same:

While a wireless cell site can provide all the computer modem, fax, pager, and phone requirements for a neighborhood—with modern fiber optic cable TV systems able to provide 200 6 MHz TV channels—what is a few kHz more for telecommunications needs? The speed of the first "cable modems" blew away the once mighty 56 kbps telephone modems. With existing access, easements, and rights-of-way, electric utilities have the same opportunities as cable and telephone companies to provide sophisticated services.

What makes a light go OFF

The tungsten filament of an incandescent lamp (flashlight bulb, turn signal lamp, or 100-Watt light bulb) changes from a very low resistance to a high resistance when voltage is applied. Tungsten has a positive coefficient of resistance for temperature. The filament of a 100-Watt light bulb changes from about 10 Ohms when "off" to a nominal 144 Ohms when energized. The 120 V ac line reaches 170 V peaks 120 times per second. (0.0042 seconds after each peak, the voltage drops to 0.) Depending on when the lamp is switched on—if the wires in the lamp, house, and those out to the pole, as well as the transformer on the pole did not limit the current with their own small resistances—the lamp could consume 2,800 Watts in an instant. This dramatic change in resistance accounts for why lamp bulbs normally burn out when they are turned on. As a bulb ages, some of the white-hot metal in the filament boils away. Some small, high intensity lamps become blackened on the inside of the glass bulb with material from the filament. As it loses substance, the filament weakens. Because of the low resistance at normal temperatures, the slight increase in resistance is not enough to offset the inrush of current when the lamp is turned on. Eventually the thermal and mechanical stresses thin and embrittle the tiny tungsten wire, causing it to break when activated.

Secret Sidebar: Ohm's law, power, and energy

This book is supposed to be in "nontechnical language." Don't let your engineering staff hear what we're about to say. The fact is, most of you deal with far more difficult areas of expertise than Ohm's law. Ohm's law describes the relationship among and between Voltage, current, resistance, power, and energy. The accounting, economic, environmental, legal, marketing, political, and regulatory aspects of electrical power generation, transmission, and delivery are all far more complex than this simple law governing electricity. It cuts both ways. Because you both use numbers in your professions, often apparently at cross purposes, much of what you do seems like an art—possibly a cursed black art—to an engineer.

An old joke reveals kernels of truth about various professions. When asked how much two and two is, a mathematician would confidently answer, "four." A scientist might respond, "all the evidence indicates four." An engineer would pause a moment and say, "just to be on the safe side, let's call it eight"—and stand there defensively, tie curling up, expecting the accountant to demand it be recalculated to three or less.

Much of what we all do is mathematical in nature. Ohm's law is much easier than compound interest. Calculating electrical energy to be generated in California and transmitted to retail customers in Nevada is far easier than developing the marketing plan and collecting revenue. Scheduling the transmission can make either of those look like child's play.

But we are generating electricity here and a basic understanding of Voltage, current, resistance, power, and energy would seem to be in order. Again, think of water (Fig. 1-3). Think of water pressure as electromotive force, measured in Vs. Think of water flowing through pipes as electric current, measured in amperes (A). Think of the size of water pipes as the resistance to the flow of current, measured in ohms (W). Think of a lake, reservoir, or water tank—if the water supply is not replenished, the flow will stop, just as electricity requires a complete circuit. Stop thinking. Are you thirsty? Just remember the limitations of analogies.

Ohm's Law

Eq. 1-1 $$I = \frac{E}{R}$$

Eq. 1-2 $$E = IxR$$

Eq. 1-3 $$R = \frac{E}{I}$$

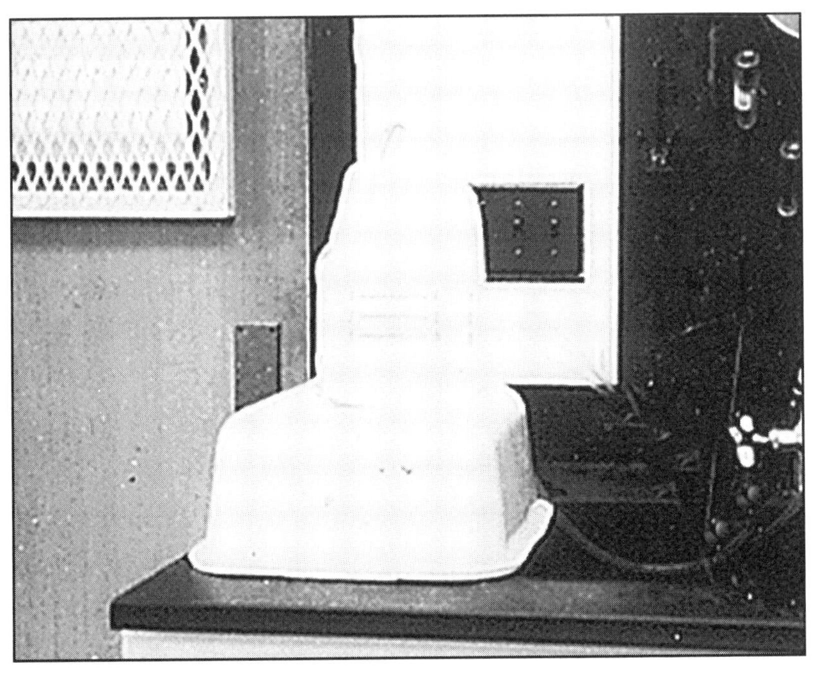

Fig. 1-3: Water Pressure and Voltage Analogy

Where I = current (flow) in Amperes (A).

Note: An Ampere is defined as a Coulomb per second of electrical charge passing through a given point in a conductor, but for practical purposes, an Ampere may be thought of as the current flowing through a circuit of 1 Ohm resistance when 1 V is applied.

E = electromotive force (electrical pressure) in Volts (V)

R = resistance (friction) in Ohms.

Power Laws

Eq. 1-4 $P = I \times E$

Eq. 1-5 $P = I \times E = \frac{E}{R} \times E = \frac{E^2}{R}$ $\qquad P = \frac{E^2}{R}$

Eq. 1-6 $P = I \times E = I \times (IxE) = I^2 R$ $\qquad P = I^2 R$

Where, P = power in Watts (W)

Energy laws, Ben Franklin, the metric system and Murphy's law

Energy is the amount of power consumed or generated, or work done during a given period of time. Energy is the rate at which power is used. Many, many different units and time periods are used by different engineers, scientists, tradesmen, etc., for many different reasons.

Notice you have not seen a nice neat energy equation yet? Recognize this one?

Eq. 1-7 $E = mc^2$

where:
E = Energy
m = mass
c = velocity of light in a vacuum

Notice the "E" was capitalized, while "m" and "c" were lower case. Notice there are no units in parentheses. Is the "E" capitalized because converting even a minuscule amount of matter into energy releases a whole pile of energy? (A pile of energy is not an actual unit; however, a ton is. A ton of air conditioning, that is.)

Everyone "knows" the speed of light is 186,000 miles per second. Scientists tend to round it off to 300,000 kilometers per second. An American engineer designing a radio frequency circuit for a UHF television transmitter might think of it as 11,803 inches per microsecond. Similarly, a computer engineer realizes it takes 1 nanosecond for a bit of data to travel 11.8 inches at the speed of light.

The contemporary metric system measure of energy is the Joule (J). A J is a "Watt-second," meaning a 1 V source causes a 1 A current (1 W) to flow in a circuit for 1 second. Even the energy used in your home in a month would result in a very large number of Joules, increasing printing costs on your electric bill, so electric energy usage is billed in kilowatt hours (kWh). Because there are 1,000 W in a kilowatt (kW) and 3,600 seconds in an hour, a kWh equals 3,600,000 J.

Speaking of the "metric system," here's a trivia question for you: Born in Boston, later sought his fortune in Philadelphia. He was the first leading proponent of the metric system in the U.S. Known as an inventor, and a ladies' man. (If George Washington was the "father of our country," this man was the father of France!) Patriot, philosopher, statesman, publisher, and scientist. (He flew a kite during a thunderstorm and defined electric current flow from "+" to "-," getting it backwards—and no wonder!) Remember, we have been resisting the metric system for a very long time. Time's up: who is Benjamin Franklin?

The basic units of electricity, power and energy—and their fractions and multiples—are inherently "metric." However, as noted above, it is often not convenient to use formal metric units for industrial quantities of electric power and energy. During this century, the metric system has evolved through three distinct versions: the centimeter-gram-second (cgs) system, the meter-kilogram-second (MKS) system, and the *System International d'Unites* (SI). The size of the base units changed by orders of magnitude between the cgs and MKS systems. The unit of length increased by a factor of 100 from the centimeter to the meter, while the unit of mass increased by a factor of 1,000 from one gram to a kilogram. In the SI metric system, the units changed little but were named after noted scientists. Thus, the unit of pressure became the kilopascal (kPa), Watt-seconds became J, and cycles per second became Hz.

It is difficult to defend 16 ounces in a pound or 5,280 feet in a mile, but there are problems with the SI metric system, too. Multiple units are capitalized (megawatts, MW), base and fractional units are not. Unfortunately, the base unit of mass is the kilogram (kg), so even though 1,000 is a rather large multiple of a gram, it isn't capitalized, and for consistency's sake, neither is a kilometer (km). Most basic and derived units are now named after famous scientists and are capitalized. Units like A, Ohm, V, and W have been retained, while other units now commemorate scientists' names.

The reciprocal of the Ohm, traditionally called the mho, is useful in calculating the resistance of parallel networks or the "negative resistance," *i.e.*, gain of electronic amplifiers; however, it adds no actual information. The SI unit for conductivity is named for the German Thomas Edison, (Ernst) Werner von Siemens. All but superconductors resist the flow of electricity. Conductivity typically indicates resistance in a medium (such as groundwater or soil) between two electrodes that are a known distance apart—normally one centimeter. The SI for conductivity, the Siemens (S), has two defects: the unit fails to incorporate the concept of resistance per centimeter and the symbol S can easily be confused with the SI for time, the second (s). The unit for resistivity that does account for distance wound up with the cute name "Ohm-meter."

The Fahrenheit (F) temperature scale has 180 degrees between the freezing and boiling points of water. "Zero" was picked by Gabriel Daniel Fahrenheit, a Dutch meteorological instrument maker (1686-1736) as the coldest temperature he could achieve with human body temperature at the arbitrary (not to mention inaccurate) 96th degree. With 1.8 degree F for every degree centigrade (or Celsius), the SI measure of temperature is too coarse even for weather reporting without tenths of degrees. Note that for SI temperature measurements, you are not supposed to say "degrees" because Celsius is the unit.

The authors' humble suggestion is for a 1,000-point metric scale between freezing and boiling to be called *milligrade*—or since it's American, MiliUS. Because people will do so anyway, make it official to use the degree (°) symbol.

So much for the metric system making things simpler. The kWh looks better and better! Murphy's law states, "If anything can go wrong, it will." A corollary to Murphy's law reveals that numerical data will always be presented in the least useful units, the classic example being velocity expressed in "furlongs per fortnight."

What would Ben Franklin say?

Chapter 2
Generation Is Energy Conversion

Some notes on terminology are in order at this point.

The terms electricity and power are often used interchangeably, as are current and even "juice". Anyone who pays an "electric bill" would be just as wise to keep their money: as shown in the section on electric conductors, individual electrons do not have to move far to conduct and the generator does not actually deliver those electrons to the customer. Similarly, in order for the electric circuit formed by the generator, transmission lines, and loads to work at all, the generator must get back *all* power it sends out to the grid. (Voltage times current measured at the terminals of the generator must equal the sum of all the voltages times the current used by loads in the network. These loads include losses in transmission lines, transformers, switches, etc.)

So what you are really paying for is electrical energy. While trying not to mislead the reader but to avoid tedium, electrical energy will necessarily be referred to as electricity, power, current, and just maybe, "juice."

Electrical generation is the generic term for converting any of a number of forms of kinetic (geothermal, hydro, solar, wind, etc.) or potential (coal, natural gas, enriched uranium, water above a dam) energy into elec-

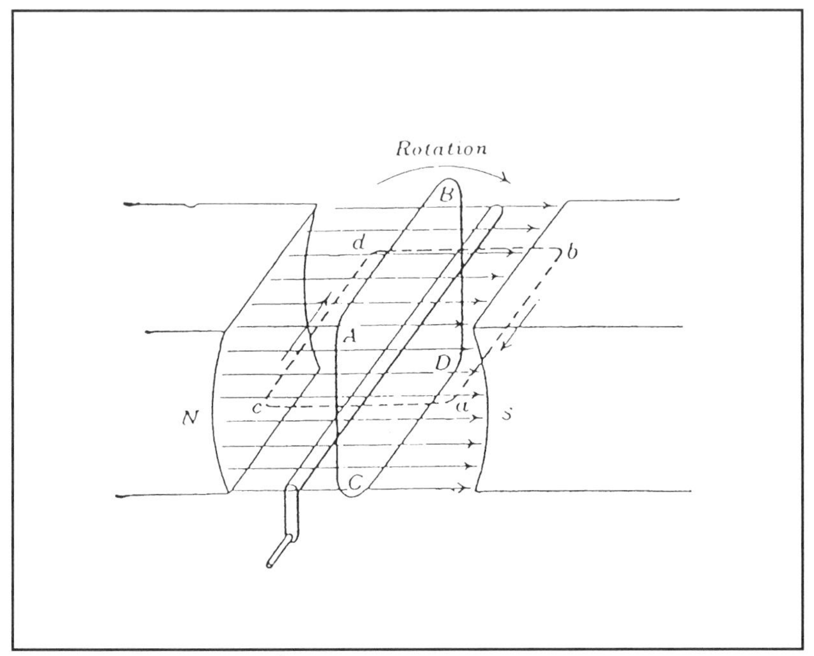

Figure 2-1: How a Power Plant Operates

trical energy. Kinetic energy is already moving, while potential energy is locked up in fossil or nuclear fuels, or water in reservoir..

Strictly speaking, just as in your car, most generators today are actually alternators.

In a traditional generator, the rotating armature coil cuts the magnetic field of the stationary (stator) permanent magnets or winding (Fig 2-1). Slip rings (ac) or a commutator (dc) conduct the induced current to the outside world.

The simpler, more reliable and durable alternator rotates the field coil and the output of the alternator is hardwired to the step-up transformer in the powerplant switchyard. In an automotive alternator, six diodes configured to rectify the 3-phase stator output provide dc for accessories, ignition, and recharging the battery. In the powerhouse, as in the car or truck, a regulator controls the current to the field coil, determining the voltage output of the alternator.

Energy conversion can and must be efficient. In the example shown in Figure 2-2, the potential energy represented by the water in the kettle when it's heated becomes kinetic energy—but without much efficiency. An

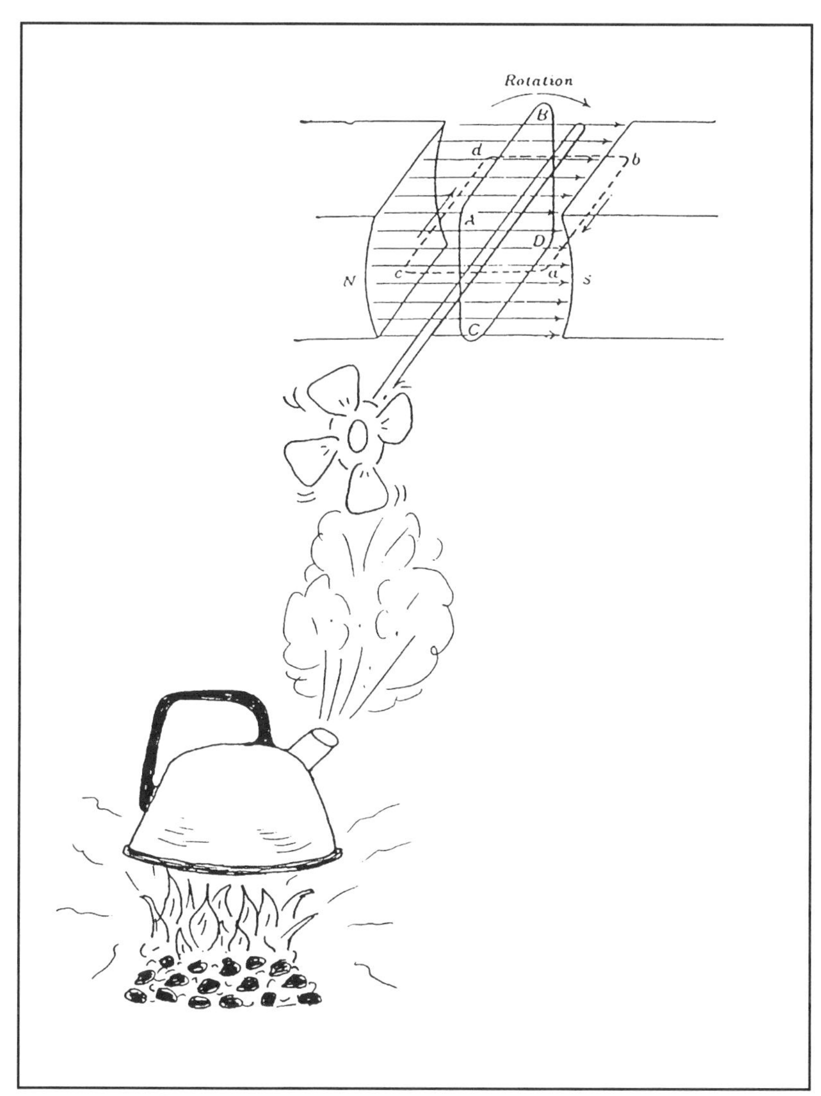

Figure 2-2: Efficiency and Power Generation

engine moves a vehicle directly via the transmission and wheels while the relatively small alternator charges the battery, energizes the ignition, and operates the accessories. In a railroad diesel-electric locomotive, a diesel engine turns a 3-phase alternator large enough to power electric motors in the truck. All three represent energy conversions, but the diesel-electric locomotives are far more efficient.

It takes energy to convert energy

Just as "It takes money to make money," it takes energy to convert energy. The gasoline engine in your car not only moves the vehicle, but also rotates the crankshaft that operates the oil and water pumps, valves and timing belt, directly or indirectly turns the radiator fan and fuel pump, and, via the alternator, energizes the computer, fuel injectors, gauges, and ignition. Once the engine is running, the alternator also charges the battery—storing enough potential chemical energy to start the vehicle the next time and powering the defroster, lights, radar detector, rear window defogger, and the stereo.

If the primary purpose of an automobile is transportation and that of an electrical generating plant is converting one form of energy into electricity, then the energy requirements of the vehicle itself would be comparable to station service in the generating plant.

Those are the useful drains on the energy produced. Exploding air and fuel inside the cylinders creates heat. Some of that heat expands the resulting gases to move the pistons, rotate the crankshaft, and so on. But only a small fraction of that heat can be utilized. Overall efficiency of an automobile engine is about 15%. Most of the energy is unavoidably wasted as heat. Fossil-fired steam plants are much more efficient, with gas turbine, combined-cycle plants nearing 60% efficiency. Hydroelectric and most alternative energy plants use "free" fuel. Nuclear steam plants are so efficient, fuel costs are relatively minor compared with the expenses of continuously and vigilantly meeting environmental, safety, and security requirements.

If your car has a manual transmission, you might get by without the battery by parking on hills or putting your passengers to work push-starting the car, but you would not want to stall in traffic.

Similarly, for startup and running, all generating plants require energy. Even hydroelectric plants need power to operate the controls, lamps, motors, pumps, relays, etc., to bring the plant on-line.

It's ALL solar

Parents lead their children into the planetarium, shushing them and trying to decide where the best seats might be with three or four together.

Soon the lights dim and majestic classical music in enhanced stereo sound fills the darkened space. As their eyes adjust, they see the oversized dumb-bell-shaped projector rise out of a central pit and spatter the domed ceiling with points of light simulating familiar constellations.

> "At the earliest moments after the Big Bang, the universe consisted of hot gases—mostly the proton nuclei of hydrogen ions—swirling through space.
>
> "The first stars formed from these protons. The shock waves of the initial concussion caused some regions of space to have slightly higher gas pressure than others. Using the word 'dense' in conjunction with the near-perfect vacuum of space seems inappropriate. However, where the wisps of hydrogen were densest, gravitational attraction began pulling them closer together—slowly at first, then ever more powerfully, inevitably forming stars and gas-giant planets. Eventually real pressures and high temperatures grew at the heart of the star being born until four protons were forced together to form a helium nucleus (two protons and two neutrons) and beginning a fusion chain reaction giving off incredible energy."

The simplified equation for the reaction is:

Eq. 2-1 $4\ {}^{1}He \Rightarrow 1\ {}^{4}He + \text{neutrino} + 1\ \text{photon}$
where:

${}^{1}H$ is a hydrogen nucleus (proton)
${}^{4}He$ is a helium nucleus (alpha particle)
1 photon is a form of electromagnetic radiation with a shorter wavelength and higher energy than X-rays, called a "gama ray."

In a second-generation star like our sun, or a more mature first generation star where helium is plentiful, a higher order fusion reaction can take place:

Eq. 2-2 $3\ ^{4}He \Rightarrow 1\ ^{12}C + 1$ photon

where:
^{12}C is a carbon nucleus

Clearly, energy systems such as geothermal, hydro, photovoltaic, and wind, are all solar. It is no real stretch to grasp that coal, natural gas, and oil fossil fuels are simply stored forms of energy derived from the sun.

So, how does nuclear power—the fissioning of uranium—qualify as solar? Given enough energy, pressure, and temperature, even the heaviest elements such as thorium and uranium are eventually formed by fusion. Young stars primarily consume hydrogen with the heavier elements being formed during the cataclysmic expansion into red giants and later contraction into white dwarfs. Fissioning of unstable isotopes of uranium, plutonium, etc., creates elements such as lead.

While these heavy elements were not created in our day star, they are remnants of previous suns, long forgotten but not gone.

Part 2: Planning and Development of Electric Power Stations

Chapter 3
Classic Considerations

In the formative years of the television broadcast industry, a "prime directive" was, "be sure there is an audience under a station's antenna." In the days of Thomas Edison's 110/220 V dc illuminating systems, customers needed to be within two miles of the dynamos. This meant that cities were electrified before rural areas. In France, the Thomson-Houston Company took advantage of the proximity of rivers to towns and cities and developed hydroelectric power. (Thomson-Houston was an American company that later merged with Edison's companies to form General Electric (GE) in the U.S. The French division became the forerunner of multinational giant Thomson-CSF, which now manufactures the GE and RCA lines of appliances and consumer electronics products.)

Today—with cable television, DBS, and Internet options available to virtually everyone—demographics are more important than geography. Similarly, with gas turbine combined-cycle generating plants economical to build and operate wherever steam is needed, and with ultra-high voltage (UHV) lines to transmit energy across hundreds of miles efficiently, electricity customers no longer need to be in the shadow of the exhaust stacks or downstream of the "prime mover."

We have touched on the changes in the electric energy industry, but it

Fig. 3-1: Oil is delivered to a conventional electrical generating plant via tanker, across the dock of a terminal constructed for this purpose. Since the 1970's, the electric power industry has turned away from oil as a plant fuel; however, plants such as this two-unit 1,000 MW facility uses oil and natural gas.

is important to remember that the overwhelming majority of generating capacity was built years ago, following the classic paradigms for siting plants. (Fig. 3-1)

Nuclear steam plants provide 10 to 25% of energy needs, depending on the area of the country. Once the wave of the future, controversy of nuclear power came to a head in 1979 with the accident at Three Mile Island in Pennsylvania and ended with the 1986 Chernobyl disaster in the Ukraine. Despite the decline of nuclear power's star, the typical U.S. "nuke" has been quietly, reliably, and safely serving the base load for 20 years.

Conventional coal-, natural gas, and oil-fired steam plants also serve base, intermediate, and peak loads depending on the size and vintage of their furnaces and generators. Hydroelectric dam and pumped storage facilities convert the energy of falling water into electric power as long as the water level is high enough to do so.

Demand for electric energy grew steadily from the 1880s, and following World War II, it at least doubled every 10 years. In the early 1970s, the Organization of Petroleum Exporting Countries (OPEC) oil embargo, dire warnings from environmentalists, economics, and world politics impacted demand. Since then, it has increased steadily but not in geometric progression (Table 3-1).

Year	Steam	Internal Combustion	Gas Turbine	Nuclear	Hydro-electric	Other	Total
1950	48.2	1.8	0	0	19.2	-	69.2
1960	128.3	2.6	0	0.4	35.8	-	167.1
1970	248.0	4.1	13.3	7.0	81.7	0.1	336.4
1980	396.6	5.2	42.5	51.8	81.7	0.9	578.6
1990	447.5	4.6	46.3	99.6	90.9	1.6	690.5

Table 3-1: Net Summer Generating Capacity (millions of kW) Source: DOE

Water

Fossil-fired steam or hydropower drove the first commercial dynamos and generators. Both require large quantities of water—the first for steam and cooling, the second as the prime mover. While the water for steam generators may be pumped out of lakes, reservoirs, or tanks as part of the station service energy requirements, obviously the hydroelectric generator must be located directly below the water supply. The greater the pressure "head" in a hydroelectric plant, the more energy may be extracted from the falling water.

While fresh water must be treated to remove debris, fish, organic matter, minerals, and rocks—material that can clog pipes, increase wear on pump seals and valves, and reduce heat transfer—saltwater presents challenges that make its use in boilers uneconomical except where it is the only choice. Salt water from Cape Cod Canal is routed to two conventional fossil-fired steam units at Southern Energy's Canal Power Generating Facility plant in Sandwich, Massachusetts. This once-through cooling system removes waste heat via heat exchangers to avoid contaminating the highly purified water used in the steam loop. The salt water cools in long,

Fig. 3-2a: Salt Water Cooling Inlet from Cape Cod Canal

open-air concrete trenches before being returned to the canal. Thermometers monitor inlet and outlet temperatures by trailing-on lines in the middle of channels.

Fig. 3-2b: Salt Water Once-Through Cooling Discharge to Canal

Cooling and steam

Where water for cooling and steam generation is plentiful, it may be used once and discharged. This saves the cost of building and maintaining high flow and volume reservoirs or water tanks. However, once-through systems limit the ability to control pH and remove minerals, increasing downtime and raising boiler maintenance costs. (Figs. 3-2a, 2b) Energy efficiency demands that as much heat as possible is extracted from the coolant; therefore, in this case, business and operating practices are consistent with good environmental stewardship.

Obviously, where water supplies are more modest, water must be recirculated. While adding to the complexity of the power plant, treating recirculating water to control pH and remove minerals that cause corrosion and scale extends the useful life of boiler and steam pipes and reduces maintenance. Even in combined-cycle plants (where steam is sold or used for heating or industrial processes) smaller diameter condensate pipes return as much water as possible from condensed steam to the powerhouse to reduce the amount of makeup water that must be treated.

Prime Mover

The term "prime mover" dates back to the Greek philosopher-scientist Aristotle, whose efforts to systematize human knowledge still represent our concept of science and the scientific method. "Prime mover" evokes images that have been used to sell bulldozers, graders and other heavy construction equipment. This black-box concept of putting diesel fuel in one end of the machine and getting progress out the other is correct in a general sense. Confusion arises when using the term technically. While a diesel-electric railroad locomotive qualifies as a prime mover in the general sense, these workhorses use a three-stage process: the diesel internal combustion engine converts potential energy in the fuel oil to rotary kinetic energy, therefore the diesel engine is the prime mover. The mechanically connected generator converts mechanical kinetic energy to electrical kinetic energy and the electric motors in the trucks convert electric energy back to mechanical energy.

For our purposes in electric power generation, the prime mover converts potential energy into rotary kinetic energy to turn the generator. Internal combustion gasoline and diesel engines are prime movers in their own right. Reciprocating steam engines, with their fireboxes and boilers, are prime movers. With fossil-fired steam turbine plants, the burner, boiler, and steam turbine comprise the prime mover. Nuclear steam turbine

Figure 3-3: Yankee Rowe Reactor Vessel Being Delivered by Rail. Photograph by Ralph K. Leach. Used by permission of Marlene Leach.

plants substitute the fission reactor for the burner, which along with the boiler and steam turbine form the prime mover. In gas turbine plants, the jet engine-on-a-stick efficiently converts potential to kinetic energy in one step and is the prime mover. In hydroelectric plants, falling water and the water turbine together constitute the prime mover.

Access

Major power plants need to be served by a navigable waterway, railroad, or road to deliver equipment and fuel, haul away ash, spent fuel, etc. Roads are nearly ubiquitous and allow employees to come and go for their shifts and permit routine pickup and delivery of equipment, materials, and supplies. Major shipping canals adjacent to generating plants assure easy fueling. Large equipment such as the nuclear reactor shown in Figure 3-3 may need to arrive by rail. When railroads were king of transportation, planning for a siding was not a major concern. Now the viability of the rail line serving the site must be considered in planning a new power station. Of course, a coal-fired plant may assure the financial success of such a line, because of daily deliveries.

Figure 3-4: Cape Cod Generation Facility's Gas Pipeline Natural gas upgrade (white pipeline) permits Canal Power Generating Facility Unit 2 to switch from oil. Plant on Cape Cod Canal (in the background) serves the Cape and southeastern Massachusetts.

Natural gas or oil pipeline

Access to a natural gas pipeline assures uninterrupted supply of an economical and environmentally preferable fuel. As part of the upgrade to the 565 MW Unit 2 of the Canal Power Generating Facility at Sandwich, Massachusetts, owners wanted to be able to use natural gas in addition to oil, so builders tunneled under Cape Cod Canal to install a gas pipeline (Fig. 3-4).

Proximity to customers

As has been noted, in the days of Thomas Edison's dc dynamos distributing power at 110/220 V dc, the customers had to be within two miles of the generating plant. The switch to ac soon began, but was not complete in the U.S. until the 1960s. With ever-higher transmission and distribution voltages, not only could more power be delivered, it could be hauled longer distances—up to 800 miles in some cases.

The value of having customers in the shadow of the smokestack is real; however, transmitting energy long distances is now a reasonable cost of doing business. Significant cost savings can be realized by upgrading shorter, existing transmission lines rather than acquiring rights-of-way, obtaining permits, and building new ones.

Favorable regulatory climate

The transition from highly regulated utilities—with a sanctioned monopoly nominally in the public interest in their service areas—to a "reregulated" competitive marketplace is far from complete. Gas turbine combined-cycle cogeneration plants finally have created such an economic incentive for reform that consumer groups, environmentalists, industry leaders, and regulators have recognized the need for change and the ability to make change occur. While the free-market approach to electric power generation sorts itself out, environmental regulations (both federal and state) march on, seemingly a "one-way pendulum."

Alternative Energy

Most forms of energy we now refer to as "alternative" were once very conventional. Only after World War II did the convenience, economy, and seemingly unlimited supply of fossil and nuclear fuels lead to building electrical generating plants that overwhelmingly outproduced other energy sources.

With six billion people in the world—between and among automobiles, existing generating plants, and heating and manufacturing applications—we could use up our fossil and nuclear energy reserves in a few years of concentrated consumption, if we really put our minds to it.

However, the effective yield of most "alternative energy" sources is limited by competing use, geography, and seasonal or time factors. For example, "Old Faithful" could be used as a source of geothermal energy but for the fact that a) environmentalists, ordinary citizens, tourists, and Yogi Bear would not stand for it; b) other forms of generation are more economical and practical for this sparsely settled area of northwestern Wyoming, and c) steam for the turbines would only be available about once an hour. While this is an extreme example, it demonstrates that harnessing alternative energy requires innovative solutions. Despite the fact the world's largest geothermal generating complex is at the Geysers north of San Francisco U.S.A., alternative energy still seems like an oddity in this country. In Iceland, geothermal energy directly heats 65% of homes.

Wind

Before rural electrification, many farms in the U.S. had small windmills to pump well water into cisterns. Intermittent winds perform this task adequately, but are of little use for generating lighting or manufacturing purposes. The gusting and wind shear caused by hilly or mountainous terrain create maintenance problems and frequent failures. As with other forms of generation, large-scale plants are more efficient. Hill and mountaintops do not have the acreage for large-scale generation. Therefore, plains with frequent, moderate, steady winds suit wind farms best.

Geothermal

Geothermal energy may only be harnessed practically where the earth's crust is thin enough to allow surface or groundwater to be superheated by molten rock (or magma) beneath the earth's surface. (Once magma reaches the surface in flows or volcanoes it is called lava.)

This suggests the biggest problem with geothermal energy: few areas of the world where geothermal energy is available are seismically or volcanically stable enough to permit its exploitation. Relatively few people live on the volcanic atolls of the Pacific Ocean because of the risk of eruptions, therefore there is neither market nor capital available to build geothermal generating plants. Iceland and some areas of the Scandinavian countries have stable geothermal resources, a market for this energy, and have invested to develop geothermal technology. U.S. geothermal electricity production has been estimated at 2.2 gigawatt (GW) for the year 2000.

Tidal

The relentless ebb and flow of tides cry out to be used as an endless source of electric energy. There are approximately two high tides and two low tides each day, depending on the relative positions of the moon and sun. Despite its incredible mass, the sun is 93 million miles away and so its effect on tides is less than half that of the moon, which is "only" 239 thousand miles distant.

During the new moon (when the moon is directly between the earth and the sun) or at full moon, about two weeks later, when the earth is between the moon and the sun, their combined gravitational pull creates the highest high tides and the lowest low tides (called spring tides). About a week after each spring tide, the moon and sun move approximately 90 degrees apart, causing minimal neap tides. For about half of each tidal period (except during neap tides) the available water flow is great enough to generate power. While coastlines zig in and zag out, multiplying their effective length, only relatively few areas make good candidates for tidal generation.

Some bays and estuaries lend themselves to damming to harness tidal energy. As another form of hydroelectric power, the environmental impact on marine breeding grounds may be just as great as damming rivers. Other suitable waterways may be navigable and the loss of shipping channels may be unacceptable.

Such "single-effect" plants open sluice gates while the tide is coming in and generate electricity when the tide goes out. "Double-effect" plants generate power while the water flows in both directions using reversible turbines similar to those in pumped storage stations. Tidal generating plants use traditional low-head Kaplan water turbines or Straflo turbines specifically designed for tidal plants.

High average solar

Cost-effective solar power generation requires large-scale solar-heated collectors driving steam turbines—and lots of sun. Desert areas as close to the equator as possible are ideal. The American Southwest comes to mind as the best compromise in the continental U.S. Because Alaska north of the Arctic Circle has high average solar during the summer, and most economic activity occurs during that period, solar power generation at Point Barrow on the Arctic coast may not be as impractical as it sounds.

Photovoltaic generation may be cost effective in some applications to

Figure 3-5: Solar Powered "Johnny"

boost sagging line voltages. However, in some cases, other costs may decide in favor of solar batteries. In the shadow of both 115 kilovolt (kV) transmission and 13.8 kV distribution lines, the photovoltaic powered "johnny" along a Cape Cod rail trail makes more of a statement about environmental awareness than energy need (Fig. 3-5). Providing 120 VAC power and phone lines to emergency call boxes along highways in remote areas would be far more expensive than installing solar powered radio-linked telephones but technology evolves so quickly that it could be cheaper yet to give away cellular telephones!

Combined-cycle gas turbines

The classic considerations for siting state-of-the-art gas turbine combined-cycle plants are not so different from fossil plants if you substitute proximity to steam customers for electric rate payers.

"Deregulation" has not and is not likely to ease the ever more stringent process required for planning and permitting construction of new power plants. Already connected by pipeline, rail, or waterway to sources of fuel and water supplies, and to customers via transmission lines, existing power stations become a valuable resource for upgrading.

Site Selection, Permitting, and NIMBY

As demand for electricity grows in the U.S., so do factors working against construction of new plants and renovation of existing ones. These include siting problems, costs, and environmental and health concerns. Much of this can be summed up in the acronym, NIMBY—"not in my back yard." This means that the residential, commercial, and industrial customers who need additional electrical service don't want new generating plants or transmission wires in their neighborhoods.

When a generator—a traditional utility or a non-utility generator—needs a new facility, it makes its best factual case to regulators. In many cases they are fighting the emotion of NIMBY. Even if a community government signs on to the utility's plan, it is often politically difficult to issue building permits. While large traditional electric generating facilities serving a wide area have always been sited in remote locations, smaller modern facilities may need to be near customers they serve. While a franchise can design a fast food restaurant to blend into almost any neighborhood, it's a lot harder with a power plant.

There are serious consequences in not standing up to NIMBY—both for the generating companies and their customers. Transmission and distribution systems that carry the electricity produced by generating plants have already experienced what can happen when new, expanded facilities are not built in a timely manner. Two major outages in the western U.S. in the summer of 1996 were complicated because of a lack of alternative transmission lines to reroute power to customers.

Site selection

A variety of factors are considered when new generating facilities are planned. Tracts of land sufficient to accommodate the plant and all its attendant systems (fuel storage and processing, substations, transmission lines, etc.) are often hard to find and expensive to buy. Transportation of fuel to the site—whether along highways, gas pipelines, or railroads—must be laid out and built.

Site selection methodologies generally involve a weighted scoring system to estimate the value of each site. Evaluations include environmental and licensing criteria as well as cost-based guidelines. Environmental criteria tend to fall into several areas including air quality, land-use planning and socio-economics, land ecology, water ecology, water quality and waste management, and geology/seismology issues.

Air quality criteria include background air quality concentrations and the potential impact of plant emissions along with projected pollutant levels and the impact the facility may have on any non-attainment areas.

Socio-economic and *land use* considerations include aesthetic value, historic and archaeological value, land use compatibility, property value of the site, public services, taxes, and transportation.

Water ecology criteria include the recreational or commercial value of the site, threats to any species, potential ecological disruptions, and special habitat such as nurseries or spawning grounds. Water quality and waste management issues include water availability, makeup water quality, soil impermeability, how close to the surface groundwater is, and potential impact to local water use and flood control.

Geological and seismological considerations are generally related to the engineering feasibility of the site. They include foundation stability of subsurface materials and groundwater characteristics.

Specialized consultants are generally used to evaluate these many complicated factors when assessing potential sites for large facilities.

Permitting

Generating plant developers must obtain federal, state, regional, and local permits. Regulations governing most permits are disseminated through federal laws that are then turned into regulations by the U.S. Environmental Protection Agency (EPA). Most of these programs have been established with intent to eventually transfer them to the states. The states and localities can implement stricter standards than those issued by the EPA but theirs cannot be more lenient. The two major areas of concern for permits are the Clean Water Act (CWA) and the Clean Air Act (CAA).

Water permits. Water regulations really began with the Water Quality Act of 1965. It was followed by the federal Water Pollution Control Act, in 1972, amended five years later. A variety of other legislation followed. All are generically referred to as parts of the CWA.

The CWA established a permit program for point-source discharges of industrial effluent into national waters and established standards so that ambient water classifications could be maintained or restored. The National Pollutant Discharge Elimination System (NPDES) identifies typical power plant wastewater streams and sets limits on what the EPA finds to be the worst pollutants in the streams. The EPA has also developed a list of toxic waste pollutants for which they are developing specific effluent limitations.

Many complicated forms have to be completed and samples have to be collected when a company applies for water permits. If permits are issued, they contain limits for emissions of conventional pollutants, plus limits for expected pollutants based upon the submitted samples. Emission limits are becoming lower as times passes and technologies enabling emission reductions improve. Once permits are issued they are generally only effective for 5 years.

Air permits. Today's air regulation began with the CAA of 1970. The Act has since been amended, most recently through the 1990 Clean Air Act Amendments (CAAA). The CAA established national ambient air quality standards (NAAQS) for major air pollutants and required states to develop state implementation plans (SIP) to maintain them. Again, states can enact stricter standards than those found in the federal act but theirs can not be more lenient. Most states have taken over administration of clean air programs from the EPA but the EPA maintains oversight authority. Regulations include definitions, standards, monitoring methods, and compliance and administration information.

The CAAA of 1990 fill almost 800 pages and 11 sections. It developed a uniform national permitting process with enforcement authority and incentives for using cleaner fuels, reducing energy waste, and implementing conservation programs. The CAAA specifically addresses the problems of urban air pollution, including smog, acid rain, and global warming.

Air permitting is the most complex of all the permitting efforts and there are specialized consultants who help companies wade through the legislation and the forms and the sampling. Different rules govern existing pollution sources and new sources. There are also differences between areas of the country classified as "in attainment" and those considered to be "in non-attainment"—the latter regions suffering more air pollution than the EPA considers acceptable. These factors, together with great many others, affect permitting efforts by power producers and transmitters.

Electromagnetic fields

For years, elements of the public have feared that electromagnetic emissions from overhead transmission lines may cause cancer, especially in children. This fear has nearly halted new construction of these lines, though ongoing debate has not determined whether, or how much, electromagnetic fields may affect anyone. A variety of scientific and medical studies are underway at all times, it seems, both in the U.S. and internationally, and one day scientists may have more definitive word. But for now we just don't know what the dangers may be.

Electric and magnetic fields appear around conductors carrying current. Because the 60 Hz power line frequency is low, the wave length is as long as the US is wide. Because power lines are balanced, the fields virtually cancel out. These fields interact with whatever is present in the immediate area. However, such fields are found everywhere in nature—in every atom. There is a natural electric field around the earth's surface created by electric charges in the atmosphere. A membrane that maintains an electric field surrounds every cell in the human body. The difference is that these phenomena are continuous while electric currents in power stations and in household appliances are made up of electrons that are displaced to create electric current. They constantly change direction (60 times a second in a 60 Hz ac). The fields from this kind of electricity are also alternating.

Electric fields surround charged conductors. The higher the voltage in an appliance, the stronger the electric field. The cord of a household appliance which is plugged in but not turned on has a slight electric field, mostly between the wires. Because of the opposing polarity between the wires, there is no net field more than a few wire diameters away from the cord. The *magnetic field* is generated by electric current—the movement of electrons. As soon as a household appliance is turned on, it generates a magnetic field. When the appliance is turned off, the magnetic field disappears. Magnetic field strength depends on the strength of the electric current and the distance from the source. The field is strongest at the source and diminishes quickly as you move away. The field from a high-voltage line falls off quickly and is no stronger than normal environmental fields at less than 100 yards.

Studies. The largest, most detailed study of electromagnetic fields effects to date looked at the possible threat posed by high-voltage transmission lines. It found no link between fields created by the lines and the most common form of childhood leukemia. The findings were based on an evaluation of 1,248 healthy and sick children under the age of 15. The study was headed by the National Cancer Institute and findings were reported in the *New England Journal of Medicine.*

Controversy over possible risks posed by power lines first arose in the 1970s. Exposure to electromagnetic fields among electrical workers had been associated with a variety of diseases, including breast and brain cancer, although there was and is no definitive evidence of such a link. Fear of these adverse reactions to fields encourage people to object to transmission and distribution lines, especially high-voltage equipment.

Childhood leukemia is the illness most closely linked to electromagnetic fields. The particular strain being studied closest is acute lymphoblastic leukemia, the most common form of childhood cancer. There are 2,000 cases of this type of leukemia reported annually in the U.S.—a 20% increase in the last two decades.

The source of the fields in these suspected cases was high voltage transmission lines rated 115 kV and up. Experts say the study cited here is a major step toward resolving the question of whether the fields raise the risk of children getting the disease. However, other experts caution that the study is not the final answer. Other studies are being conducted and may shed more light on the subject. Three earlier studies indicated that living near electormagnetic fields created from power lines and heavy electrical equipment might slightly increase the risk of contracting acute lymphoblastic leukemia. However, these previous studies were small and had scientific problems that the latest study addressed.

Chapter 4
Technology-driven Changes

In a move reminiscent of a Monty Python sketch, Western Massachusetts Electric Company (WMEC), owned by Northeast Utilities, recently announced it was selling most of its generating capacity in Massachusetts to a new subsidiary it would set up for the purpose. Stories like this are common as the electric power industry moves from highly regulated monopolistic utilities to a marketplace reregulated to promote competition. What appears to be a shell game to cynical consumers means survival to traditional utilities and newcomers alike in an industry driven by technological change.

The single biggest technological change driving the electric power industry is the economy of the gas turbine combined-cycle cogenerator. Previously, economy of scale could only be achieved using huge fossil or nuclear steam turbine plants. A gas turbine generator is essentially a jet engine on the same shaft as the electric generator, cutting out the "middle man" boiler between furnace and generator. But as they say on the Ronco TV commercials, "there's more!" The exhaust gases from the gas turbine are so hot they can still boil water to run a conventional steam turbine generator. ("How much would you expect to pay for both of these benefits?

Wait, there's still more!") The exhaust of the steam turbine generator is superheated steam usable for district steam heating or industrial processes. ("How much would you pay for gas turbine generated electricity, plus steam turbine generation, plus steam for heating or manufacturing?")

Well, it is not $19.95 plus shipping and handling, but an industry that was burning fuel to make steam anyway can now get electricity for its own use and to sell for the same cost as the fuel plus the construction and operating costs of the generation plant. Where possible, utilities have always sold steam from generating plants. Now, when demand rises, new economical gas turbine combined-cycle cogenerating plants deliver more kW hours economically and still have steam for sale.

While gas turbine combined-cycle cogeneration is the single biggest technical factor within the electric power industry driving change, other technical and social changes contribute to the accelerating changes. A quadruple win situation—for consumers, the corporation that owns the plant, the environment, and the steam customers—the gas turbine exemplifies how technology is driving changes in the electrical energy industry.

The role of technology

At least in the short term, electricity could still be generated, transmitted, and distributed without computers. Simple analog meters permit generating plant personnel to control power factor and voltage and to synchronize the ac waveform with the grid. Digital and computerized controls and monitoring equipment have crept into a nuclear power plant's control room, but the original gauges still in everyday use are conventional analog. However, as with every other aspect of our lives, ever more powerful computers and computer networks, including the Internet, determine how utilities and unregulated generators acquire capital, bill customers, market electric energy and schedule its delivery.

According to Moore's "law," computing power has doubled about every 18 months since before the monster vacuum tube ENIAC, EDVAC, and UNIVAC computers revolutionized the computer industry in the mid '40s to late '50s. This guideline, formulated by Gordon Moore, co-founder of Intel, is expected to hold through 2020. Then, with the development of quantum electronic and biological computers, the rate of change may well increase.

And technology changes spring from—and drive—economic and political change. Free enterprise generates capital for investment in new technology. The downfall of Marxism and Communist governments in countries around the world in the late 1980s and early 1990s, while democratic countries with entrepreneurial economies enjoyed growth and prosperity, led to a renewed belief in the value—and inevitability—of the free market economy. Suddenly, Adam Smith's "invisible hand" could be seen everywhere. Countries such as Russia, which systematically decapitalized their economies following "ze Revolution," are still trying to come to the party—the free enterprise party. In the West, consumers, corporations, and regulators embraced deregulation—more properly called "reregulation"—of the electric power industry based on the success of the breakup of the telecommunications industry and the sustained prosperity and low unemployment rates anticipated through the Millennium, perceived to be the result of freer enterprise.

New technologies replace older technologies, and demand causes money to move from one supply of goods or services to another. More often, the new product or service creates a new demand or fulfills an unmet demand. While McDonald's did put many diners and "greasy spoon" restaurants out of business, many diners, luncheonettes, and fine dining restaurants found they could compete quite successfully with fast food establishments because they provided a more complete menu, a more relaxed atmosphere, personal service, a second cup of coffee, tastier food—whatever.

Today, the technological change with which we reckon—the gas turbine cogeneration plant—will spawn distributed generation and merchant plants because of market changes pioneered by the gas turbine plants. Getting high temperature superconductors out of the laboratory would revolutionize the transmission of electric energy, encouraging more "mine-mouth" coal generating plants. With nuclear plants able to be sited where they are environmentally and geologically safe and secure from terrorist threats, with good on-site storage solutions and the possibility of a hub-and-spoke arrangement of power and breeder reactors, a financially viable and well regulated nuclear power industry could conceivably be re-born.

Distributed Generation: Bringing Power to the People

Distributed generation is emerging as a cost-effective way to deliver electric power when small loads are needed close to the end user. Generating facilities providing tens of kW or MW to satisfy local demand are used in place of huge central generating stations and attendant transmission and distribution networks.

Such solutions are becoming available because:

- proprietary, large-scale plant design and construction methods allow for high volume, assembly-line production of generation equipment
- deregulation and market competition are fostering innovation
- technology advances are promoting choice

These advances include individual solar photovoltaic arrays for residential power applications, microturbines for commercial business needs, and reciprocating engines for industrial cogeneration applications. "Peaking" capacity, power quality, and other requirements can be supplied with fuel cells, aeroderivative and industrial gas and wind turbines, and battery and flywheel systems. These are just the most common technologies at this writing.

Existing capacity is already being utilized in certain instances as distributed generation. Facilities may deploy back-up generators as "peak shavers" in addition to emergency reserves. Control and dispatch software, still under development, will enable standby generators to be connected to the grid profitably.

Distributed generation will never completely replace centralized power stations, nor does it have to be cost-competitive with them. Instead, its specific customer benefits must outweigh any additional costs. In fact, end users need not be the only market. As suggested above, grid managers can use distributed generation as a part of a strategy to maintain network integrity.

Distributed generation is as simple or as complicated as a user cares or needs to make it. Facilities of between 5 and 100 MW are sized to serve individual customers or "clusters," neighborhoods, towns, and industrial parks. They serve as non-base load generation to aid utilities with peak loading. Conversely, they can be considered totally "off-grid base load" that uses the utility and transmission system as the back-up. Generator sets installed at customer facilities, operated during peak utility-demand periods, enable utilities to cost-effectively meet demand and efficiently use generation/transmission systems. They also enable customers to avoid load curtailment (and gain cost savings) and, during peak demand periods, avoid peak power consumption demand charges.

Additional savings come in time. Planning and building a large power plant averages between 5 and 7 years, with many more years required to pay off the investment. Distributed generation facilities can be up and running in a year's time (or less) and paid for in five. Gas turbines are the fastest—another plus for them. Siemens markets a turnkey operation that's up and operating in 18 months. Both utilities and customers benefit when capital is freed in this way. In addition, utilities gain capacity without having to invest in costly new generating plants or transmission and distribution upgrades. Customers potentially receive more reliable service, lower bills, and combined heat and power.

In an even larger economic sense, distributed generation is one of a number of alternatives regulators and electric utilities are investigating and investing in as the industry is deregulated and privatized, and options become available to industrial, commercial, and residential customers. Many of them will select to continue a relationship with an existing electricity provider that supplies power from the grid, while others will opt for on-site or local generation.

"Is distributed generation for you?" is what power marketers and independent generators need to ask current and potential customers—and then be able to explain why distributed generation offers specific and measurable benefits:

- the ability to "self-generate" when grid electricity is expensive, thereby reducing costs
- the ability to provide power to the grid when revenue opportunities are high
- the ability to generate steam, heat, and/or power for process needs
- the ability to maintain power to critical loads when grid power is unavailable, avoiding lost production, lost information, and lost customers

These benefits are available to customers who have established:

- defined load use patterns
- fuel costs (including all competing fuels)
- transmission and distribution system reliability and avoided costs
- available technology
- up-front capital costs
- annual equipment maintenance costs
- purchased electricity costs (for comparison)

Efficient use of distributed generation is also dependent upon where customers live or operate a commercial or industrial facility. In the U.S., it is most popular where energy costs are highest—the east and west coasts, Texas (because of demand charges), and where growth is causing transmission and distribution bottlenecks.

In other developed countries, distributed generation similarly represents an alternative to conventional electricity supply and/or contribution to the grid. In the developing world, where millions lack access to electricity—where "off-grid" is the rule and not the exception—distributed generation represents an opportunity to increase the quality of life without waiting for construction of central-station power plants and transmission and distribution systems.

Since the genie has been let out of the bottle by de/reregulation, distributed generation, like any technology driven change, is more a matter of the process of "awareness, acceptance, action" than any specific engineering development or scientific discovery. Gas turbine generators have been around for half a century. It is not so much the technology, but the realization of what we can do with the technology that is revolutionizing electric power generation.

History

The most often used analogy for distributed generation vs. traditional power plants is the evolution of the computer industry. Just as centralized mainframes were supplanted by PCs, so the electricity industry may devolve from centralized power plants to distributed power resources.

An important perspective to keep in mind from the computer industry is the periodic swings from central computing to distributed computing and back. The original mainframe computers used hundreds of thousands of vacuum tubes and essentially required their own power plants. Now five million or more transistors can fit in a two square inch microprocessor and a notebook computer can run for several hours on its internal battery. But because "the more things change, the more they stay the same," most business computers are again linked via networks to central servers. Connectivity makes everything better as no one computer has all the data necessary to do most jobs and afterwards the results must be shared to be useful.

The other analogy comes from the earliest days of the electric power industry itself: Edison's Pearl Street station can be thought of as a distributed generation facility, if only because of the limited reach of his dc product. In fact, Edison's example proves that distributed generation was "there first." Lacking interconnected transmission and distribution networks, cities were on their own to generate and distribute electricity. Rural end users seeking electricity's benefits had to generate it themselves. (Farmers, for example, used their windmills to power their pumps and lights.) As transmission networks finally developed and grew, distributed generation succumbed to the convenience (and lower cost) of grid-supplied, central-station generated electricity.

After Edison's Pearl Street model was replaced by Westinghouse's central-station model, conventional wisdom and practical applications continued to flow towards the "economies of scale" approach for new power plant construction. Bigger plants meant lower unit costs ($/kW). In their final incarnation, in the 1970s and 80s, fossil-fired units were built with capacities greater than 1,000 MW. For high-load growth conditions (7–10% per year), investment in such large units seems justified. As load growth slowed, however, payback for such investments stretched out too far as demand wavered. The emergence of the independent power producer and development of highly efficient generation technologies for smaller loads has resulted in a growing awareness and interest in competitive, smaller scale generating plants.

As utilities formed grids to share excess generating capacity and provide for emergencies, it quickly became evident that demand was outstripping capacity—and that nobody really had "excess" capacity. Everyone had to buy power to meet demand, spurring development of new power stations and technology within the regulated framework.

Technology changes are always accompanied by an evolution in industry practices. Centralized, command-and-control regulation is becoming competitive and market driven. Countries privatizing their electricity industries are following the U.S. model. As generation is transferred from government or utility ownership to private hands, efficient plant management and low-cost operations become critically important. Generators must assemble and maintain the most effective combination of central station and distributed resources. Competition evolving from privatization will allow consumers to select generation alternatives. Distributed generation will benefit both from utilities that invest in it to retain key customer loads and from customers that invest in it for self-generation or for combined heat and power.

At this writing, it is predicted that up to 30% of new generating capacity will be met with distributed generation in the next decades. Overseas, combined heat and power plants are increasing their use of it and village power electrification programs are popping up throughout Africa, Asia, and South America.

Technology options

As mentioned above, distributed generation refers to generating assets between 25 kW and 50 MW. This includes aeroderivative (literally, redesigned aircraft jet engines) gas turbines, microturbines, reciprocating engines, fuel cells, wind turbines, photovoltaics, batteries, and flywheel systems. Initial cost favors turbines and engines but their efficiency trails the other alternatives for many distributed applications (small capacity needs) and because of certain operating problems (noise, emissions, and maintenance). Renewable fuel-based generation, batteries, and flywheels cost more but may satisfy power quality, voltage support, and remote power needs.

Cogeneration—combined heat and power—is an "inside-the-fence" application for distributed generation for industrial sites needing electricity and steam and cities that need electricity and hot water.

Chapter 5
Fuels and Fuel Handling

At it has for many years and will for the foreseeable future, coal generates the majority of electric power in the U.S. A distant second, nuclear fuels generate about a third as much electricity as coal, and half again as much as petroleum and natural gas combined. Hydroelectric produces a little more than half as much as nuclear power and is gaining on it, especially as nuclear plants reach their end of life and are decommissioned. Natural gas and hydroelectric compete for third place, with gas probably the short- and middle-term winner. (Some day we will run out of natural gas, but unless "global warming" gets completely out of hand, there will still be water.) Petroleum generation ranks a solid fifth. Lumped in sixth place, all other "renewable" energy sources produce about 10% as much electric power as petroleum.

Together, fossil fuels provide the overwhelming majority of power in the U.S. Grouped as fossil fuels, nuclear, hydroelectric, and renewable, the top three categories of fuels would not change without gas or oil (Table 5-1).

Year	Coal	Nuclear	Hydroelectric	Gas	Petroleum	Renewable
1994	1,635,493	640,440	243,693	291,115	91,039	8,993
1995	1,652,914	673,402	293,653	307,306	60,844	6,409
1997	1,787,806	628,644	337,234	283,625	77,753	7,462
1998	1,807,480	673,702	304,403	309,222	110,158	7,206

Table 5-1: 1994-1998 Net Generation from U.S. Electric Utilities by Energy Source (million kW hours). For complete 1997-98 numbers by state, see Appendix A (Source: EIA).

Looking only at the benefits of any one fuel, one might be convinced it was the ideal energy source, to the exclusion of others. With our coal reserves, the U.S. could be energy self-sufficient for the foreseeable future. Modern strip mining methods are highly productive and safe as any construction site. Trains efficiently deliver coal to power plants.

With gas turbine combined-cycle cogeneration plants cost effective to build and operate, what could be better than being able to put power plants with "smokeless" smokestacks anywhere you need steam? With nuclear power plants, the cost of fuel is negligible relative to other expenses. Several hundred tons of fuel rods will operate the plant in its design life, no black snakes of railroad hopper cars required. During normal operation, there are no hazardous emissions from nuclear plants, certainly no CO_2 or other "NO_Xious" greenhouse gases to contribute to global warming. (Fig. 5-1) Using water as a "fuel," hydroelectric plants take in and discharge water, producing electricity for free. Fish ladders allow salmon to swim upstream, spawn and die; how could anyone ever consider anything other than hydroelectric power?! In the real world, each fuel has its challenges, strengths, and handling requirements.

When you pull up to the gasoline dispenser to fuel your car, your choices are limited to three, perhaps four grades that differ from one another in their octane and perhaps the additives used to keep engines clean. That's it! Those who say, "gasoline is just gasoline" are mostly right. However, when it comes to fuels for electric power plants, the differences are very real, and the choices made by plant owners and operators depend on many factors. Some of them are dictated by state and federal regulators or even the communities in which they locate; some of them are made for economic and environmental reasons. In this case, "fuel is not just fuel!"

Fig 5-1: NO_X Control Steam Line

The fuels used to fire the boilers of electric power plants are divided into two broad groups. Traditional fuels include coal, natural gas, and oil, as well as the uranium that allows nuclear power plants to operate. These are the leading fuels in U.S. power generation. They differ from renewable fuels in that once coal and uranium are dug from the ground, and gas and oil pumped out of reservoirs and burned—they're gone. They cannot be replaced. The renewables—primarily wind, solar, water, the heat of the earth's core, and even man-made trash—have the ability to replace themselves.

Coal remains *numero uno* among generators, and is expected to hold the crown for several decades, despite the rise in use of natural gas, especially in new construction and retrofitted units. Nuclear energy currently generates about 15% to 20% of U.S. power needs and its market share is expected to drop as the economics—and the politics—associated with it continue to suffer. Oil trails with less than 10% of the electric power market.

Given the occasional uproar concerning use of nonrenewable natural resources, why do fossil fuels remain popular? It's because there are a lot of them, and they're economical to mine and use. Look at it from the generator's point of view: his fuel costs are the largest part of his maintenance and operation costs, so every dollar he saves is money he's earned. Currently, fossil fuels account for about 80% of total U.S. energy production and that production consumes about 85% of their recoverable supply. The infrastructure needed to move them from the mines and the oil and gas fields to the generating plants is in place and we've known from the industry's origins how to use them.

Changes are occurring within the fossil fuel world, however. As was noted, much of the new generating capacity currently being installed is natural gas, for the simple reason that it burns cleaner than coal and so eliminates or reduces many emissions issues. Natural gas is economical and new technologies ensure that these plants are at least as efficient as those burning coal and other fuels—if not more so.

Traditional Fuels: Coal, Natural Gas, Nuclear, and Oil

Coal: the U.S. version of Middle Eastern oil

Coal remains the fuel of choice for U.S. electricity generation. It is burned in 55% of our electric power plants. Relatively inexpensive and reliably available, it serves the U.S. as a fuel resource the way oil serves the countries of the Middle East and OPEC. We have more coal inside our borders than any other country in the world—probably 24% of the world total. Recoverable amounts are found in 38 states and make up about 95% of the nation's fossil energy reserves. Total U.S. coal resources are estimated at about 4 trillion tons.

Generating electricity consumes 8 of every 10 tons of coal mined in the U.S. The 500 power plants in the U.S. that burn coal can use up to 20,000 tons in a day. Forty-four states have coal-fired power plants within their borders.

That's a pretty tall order for an ugly, black (or brown) rock that contains sufficient organic matter and carbon (along with other elements and minerals in varying degrees) to be able to be burned. The gold inside the dross is coal's Btu value—its ability to burn hot enough to fire the boilers in generating plants. (Btu is an acronym for "British thermal unit"—the amount of heat required to raise the temperature of one pound of water 1 degree F.)

Coal begins its long development life as peat, a brownish-black organic matter that resembles decayed wood. Through generations, the pressure of accumulated layers of sediment and rock turn peat into seams of coal imbedded in other types of rock such as shale, clay, limestone, or sandstone. The more pressure and stress brought to bear upon the peat, the higher the grade of coal that results. People in some countries of the world use peat as a low-grade fuel for residential heating and cooking, but in the U.S. it's more often found in gardening uses as peat moss.

Four additional and distinct types of coal are recognized in industry: lignite, a brownish-black coal with high moisture and sediment content and the lowest carbon content and heating value (it's considered the lowest quality fuel); subbituminous coal, dull black and with a higher heating value than lignite; bituminous coal, a soft, intermediate grade of coal most common and widely used in the U.S.; and anthracite, the hardest type of coal, nearly pure carbon. Its heating value is the highest.

Coal is mined via tunnels dug into the earth (underground mining) or by removing the earth that covers it (surface mining). Though we may picture coal mining to be practiced the pick-and-shovel way, as it was in the film, *Coal Miner's Daughter*, it is actually a complex, technologically advanced operation that uses sophisticated equipment. It's getting more efficient, too: in 1945, 383,100 U.S. coal miners averaged almost 6 tons of coal production daily; 50 years later, 95,700 miners averaged more than 40 tons daily.

Concerns have been raised in recent years about coal mining. As a result, once coal is removed from surface mines, the overburden is put back, regraded, and replanted so that the land can be reclaimed. Underground mining is less intrusive: shafts dug down into the earth do not disturb the surface of the mining area.

Once brought to the surface, coal is inspected, cleaned, and prepared for use. Its quality and chemical makeup varies greatly (even when it's mined from the same seam) so coal is showered and graded to separate higher quality from lower-grade product and to remove clay, rock, or shale. (Some 30 tons of refuse is removed from every 100 tons of raw coal that is cleaned.) Prepared coal is stockpiled and shipped to customers, usually by either railroad or barge. Upon delivery to the customer, it is pulverized into fine particles and injected into a boiler's burner. Each ton that a power plant consumes generates about 2,000 kWh of electricity.

Where coal draws less than rave reviews concerns its emissions. Coal, when burned, generates more greenhouse gases (air pollutants that the government wants to eliminate) than any other fuel. Generators are under pressure to cut emissions of these gases—pressure that often equals that suffered by peat, as it evolves into seams of coal! Flue gas desulfurization units have been developed to enable coal-burning plants to comply with emissions limits, but installing and maintaining such equipment and training personnel to use it impacts a plant's bottom line. Often this variable makes other fuels or technologies more economic for a utility that must carefully watch costs. It's this variable that's driving the switch to natural gas as the fuel of the future in electric power generation.

Is Coal Still Elvis?

> "King Coal rules West Virginia like a petulant monarch, one used to getting its way."
>
> —Maryanne Vollers
> "Razing Appalachia"
> *Mother Jones*, July 1 1999

For most of this last century, coal was king—king of the energy industry. It fueled and defined the pace of America's growth. Despite challenges from environmentalists and technological advances to other fuels (most notably, natural gas) the biggest challenge facing coal may lie in the emerging competitive market for electricity and the restructuring of the electric industry.

Currently, coal fires 55% of all electric power generated in the U.S. This is equivalent to 2.5 million tons per day, generating 7.7 million MW hours per year. Forty-five percent of all U.S. generating capacity is coal-fired and four out of every six railroad cars coming down the track carry coal.

While the increasing efficiencies of natural gas (especially in combined-cycle plants) beat coal in the quest to fuel new power plants, state-by-state deregulation and industry restructuring are changing the way fuels are purchased and electricity is generated. Since generation accounts for 88% of domestic coal consumption, these fundamental changes will change coal markets more profoundly than any challenges ever before faced.

Some of the issues facing competitive coal markets are listed below:

- How are mergers, acquisitions, divestitures, and the sale of generation facilities directly impacting coal sales? The overarching issues here are that utilities (and the new "generation companies") must drive down energy costs in an industry that was built upon protected profits, fixed costs, long-term contracts, and stable production.
- As generating facilities change hands and new ones are built, how can coal retain and grow its markets?
- Coal-burning plants have always topped environmentalists' lists of alleged air polluters. New environmental initiatives on the national, state, and local levels could impact coal producers and users. Topping the list of concerns: proposed EPA rules regarding control of mercury, ozone production, and "fine" particulate matter; ozone "transport" and regional air quality (the alleged transport of "acid rain" from Midwestern power plants to the Northeast); emissions "trading;" and the Kyoto proposals for reducing greenhouse gases.
- Despite the U.S. having more coal than any other country, certain market factors spurred a dramatic decline of exports and an increase in coal imports during the first half of 1998.

- Coal transportation has generated several issues:
  - √ Fifty million tons a year are barged throughout the Ohio River system to power plants along the waterway. How volatile is this market?
  - √ If CONRAIL routes are split between the CSX and the Norfolk Southern systems, how would this alignment affect coal distribution? What other "rail issues" are out there?
  - √ What is the status of technology changes, such as transportation through pipelines ("coal slurry")?
- Coal markets are coming under the influence of futures trading and price hedging strategies that have greatly influenced other energy product markets

Buying, selling, and trading coal with such financial instruments could enable buyers and sellers to hedge price risks associated with the increasingly volatile coal market. While this reassures financial markets, it imposes changes on the traditional buyers and sellers of coal. What had been considered virtues—steady pricing and long-term, fixed-price contracts and single-sourcing—is considered undesirable in an energy industry becoming increasingly sophisticated in its ability to trade Btus, not just physical fuel supplies.

One example: coal tolling. An energy marketer pays a utility to burn a quantity of coal and take the power generated. In reverse tolling, he sells generation to a utility to gain rights to its coal, which he then sells on the open market or transfers to another competitive plant.

These issues equate to several realities for the coal industry as hard as anthracite. The market for new base load generation—the historic source of strength of coal—will continue to decline as industry restructuring frees excess capacity to be repackaged and sold by power marketers. As the aging "coal fleet" needs to be replaced, plant developers will be drawn to the lower costs, lower capital outlays and technological efficiencies of combined-cycle natural gas plants.

The bright spot: the U.S. wholesale electric market is evolving to a market-clearing price for power, and because low cost coal plants are largely depreciated, they can compete effectively on marginal price.

Fig 5-2: Gas turbine generator in operation: No one can hear it scream.

Natural gas

Deep underground—far deeper than coal—we find natural gas. It's a combustible, gaseous mixture of simple hydrocarbon compounds composed almost entirely of methane, CH_4 with small amounts of other gases including ethane, propane, butane, and pentane. It has been used for many years for home heating and cooking though it was first used in the U.S. in 1816 for streetlights. Its use as a fuel for electric power generation is increasing, and for the same reason: it's the cleanest burning of the fossil fuels, producing only CO_2, water vapor, and small amounts of NO_x. (Fig. 5-2)

Natural gas was formed when tiny sea animals were buried beneath sand, mud, and rock. Layers of organic and inorganic matter built up until pressure and the earth's heat turned them into petroleum products—including natural gas. This explains why natural gas and oil are often found together. It is sold to purchasers in measurements of thousands of cubic feet (Mcf) from wells in production fields, though consumers are billed by heat content (Btus). A cubic foot of natural gas has about 1,027 Btus.

Currently, though only about 15% of U.S. natural gas supplies are used to generate electricity, that percentage is growing, as has been noted, and new technologies coming into play will ensure this trend continues. Combined-cycle systems use waste heat to produce more electricity, while cogeneration uses waste heat to provide heat or steam to industrial processes. In addition, environmentalists are happier with gas than with coal, as emissions are fewer and natural gas-fired generating units are cleaner. They produce less than 1% of the SO_2 and particulate emissions of coal, and 85% less NO_x than a coal plant fitted with pollution control equipment.

Nuclear

Nuclear power plants have a lot going for them in terms of reliability, efficiency, and a lack of emissions. The drawbacks are all fuel related.

Nuclear power was first used for electric generation in the U.S. in the mid-1950s, when a commercial reactor was built in Pennsylvania as a cooperative effort between the U.S. government and industry. The industry took over construction and operation of nuclear plants thereafter, reaching their glory days in the mid-1970s when 109 reactors were in service.

However, a series of accidents—capped by the 1979 incident at GPU's Three-Mile Island (TMI), Pennsylvania plant and the Chernobyl plant in Russia, in 1986—caused them to lose popularity and, eventually, financial credibility. Following TMI, orders for nuclear reactors were cancelled and another U.S. plant has never been built. Nuclear power remains popular overseas, particularly in France and Japan, which still have active nuclear generating bases and construction plans. Interestingly, even in the absence of new plants in the last two decades, nuclear power continues to meet 40% of the U.S. demand for electricity.

Nuclear power plants generate electricity the same way fossil plants do—a boiler produces steam that spins a turbine that generates electricity. The major difference is the fuel—nuclear plants use uranium instead of coal, natural gas, or oil. (Actually, what's used is enriched uranium: most naturally occurring uranium is ^{238}U; the ^{235}U needed for a nuclear chain reaction can be found less than 1% of the time. To make it, the strength of the ^{235}U is boosted about 3%.) The fuel is made into pellets and installed into tubes called fuel rods. The rods are bundled into a fuel core.

Most nukes are described as light water reactors (LWR), with the fuel core submerged in water to slow the neutrons during fission and to conduct heat away from the reaction. LWRs can be either boiling water (BWR) or pressurized water (PWR) reactors.

Moreover, nuclear power retains its primary advantage—a lack of harmful air emissions: nuclear plants produce no greenhouse gases at all. Throughout the last 20 years, they have produced electricity without the emission of 80 million tons of SO_2 and more than 30 million tons of NO_x that would have accompanied generation via other energy sources. Nukes have been responsible for 90% of all emission reductions by the electric utility industry since 1973, according to industry figures. Average industry-wide capacity is above 70%, up 16% from 1980, mostly due to plant modifications, improved operating and maintenance practices, and better personnel training.

Emotionalism aside, nuclear energy has proven itself to be competitive with all other sources of electricity, with its average production costs (including fuel) only marginally more expensive than coal and less expensive than natural gas. The major emissions problem has to do with spent fuel: it's radioactive and difficult to safely dispose of. Yet, radiation is a natural part of our environment. It's in the rocks and soil all around us—even we humans are slightly radioactive—and man-made radiation accounts for only about 18% of the total. The radiation produced by nuclear power is called ionizing radiation (possessing sufficient energy to knock electrons out of atoms, producing ions. Non-ionizing radiation is the kind found in radio waves and television transmissions). Radiation produced by the operation of a nuclear reactor includes neutrons, alpha rays, beta rays, and gamma rays.

What's more, only a small fraction of the nuclear waste is highly radioactive. A 1,000 MW reactor produces about 30 metric tons of spent fuel annually. Some of the radioactive isotopes decay within a few hours, days, or weeks, while others remain radioactive for centuries.

The federal government is formally charged with managing disposal of spent nuclear fuel, but despite mandates from Congress in 1982 and 1987, it has defaulted on its promise to transfer spent fuel from plants to a permanent storage facility (the latest unmet start date was Jan. 31, 1998). Meanwhile, storage facilities on-site are filling. The final chapter on this saga has yet to be written.

The other problem nuclear plants suffer is the expense involved in building them, when compared to fossil fuel plants. Much of this added expense stems from required safety features and the lack of standardized designs, which would lower construction costs and shorten construction time. Nuclear power plants are regulated by the Nuclear Regulatory Commission (NRC) and the EPA. Their strict rules regarding worker safety and waste treatment are stiffer than those under which fossil plants operate, to protect people living around the plants as well as those working in them, and to protect the environment from nuclear exposure.

Another added expense for nuclear plants is their eventual decommissioning—taking them out of service. After an expected lifetime of between 30 and 40 years, plants are expected to be removed from service. Part of what electric customers pay to generators is put aside to pay for this—costs of probably several hundred million dollars! Add this to the expense of building in all the redundant safety features, and you see why nuclear power is considered by many to be uneconomical in a deregulated environment where cost containment cannot be overlooked.

Ultimately, the bottom line for nukes is the bottom line. Before the onset of market deregulation, generators were permitted to recoup operating costs plus a "reasonable" profit. In a deregulated market, they'll only be able to charge the market rates for their product. Operators of comparatively inexpensive natural gas or coal plants are expected to be able to underbid the nuclear generators, although fuel costs for natural gas and coal plants are higher. The nuclear plants, alas, have larger and unrecoverable capital expenses.

Oil

Oil is no longer the fuel of choice for power generation (accounting for only about 10% of domestic production). It is more expensive to use than natural gas and dirtier, and so doesn't measure up to increasing environmental regulations. It's also un-American, in a manner of speaking—most coal and natural gas is produced here (though a portion of our natural gas comes from Canada) while oil is largely imported, much of it from the OPEC countries.

Oil quality—and cost—are other issues. Crude oil is 82 to 87% carbon and up to 15% hydrogen with the remainder consisting of sulfur, nitrogen, and oxygen. The higher the sulfur count (the more "sour" the crude), the lower the quality and the higher the pollutant levels; the lower the sulfur levels (the "sweeter" the crude), the less polluting it is, but the cost is significantly higher.

Alternative fuels: hydroelectric, solar, wind, and biomass

Though discussed more thoroughly in chapter 15, we need to at least mention these fuels here to set the stage.

Coal, natural gas, nuclear fuel (uranium), and oil are examples of fuels that are nonrenewable—burn them once and they're gone (except for the residual matter). Renewable energy sources are those that will replenish themselves—things like the tides, water-driven mill wheels, the wind, and the sun.

For decades, these fuels have been eyed with regard to producing electricity, but their technologies were such that they were too expensive and inefficient vis-a-vis traditional fuels discussed above. However, this is changing on several fronts. Movement away from large, centralized power plants means small generating stations may finally make sense—the kinds of applications for which these fuels can efficiently and economically compete. Restrictions on air pollutants, advancing research and development, and (in some cases) government subsidies, are enabling these alternatives to be able to compete.

Renewables' public image also serves them well. By offering them as examples of "green power," several utilities have been able to sell customers on image as well as product. In addition, because renewable energy is available domestically, it is safer and "more American," perhaps, than using imported fossil fuels.

However, renewable fuels still operate in a niche market. Traditional alternative fuels—solar, wind, geothermal, hydroelectric, and biomass, as well as municipal solid waste (MSW)—provide only 12% of the U.S electricity supply at this writing. Of this total, hydroelectric provides almost

10%, while biomass and MSW together contribute only 1%. What's more, the growth of hydroelectric is slowing as a lack of additional large sites limits its potential.

The other members of the renewable family—geothermal, wind, and solar—together make up less than 1%. The term "green" can also refer to their history—most all are relatively new to the electric power market. Geothermal, wind, solar, and MSW resources have only been around since the 1980s, and only because of a combination of technological improvements, governmental encouragement (*i.e.*, money), and the increasing costs of using fossil and nuclear fuels.

Assuming, that hydroelectric power holds the lion's share of this market, how are the others faring? For some—wind and solar thermal generation—advancing technology will translate into improvements in generating costs and may increase their market penetration. For solar (photovoltaic) and geothermal, large cost reductions will be necessary before greater market penetration occurs.

Wind

Picture in your mind a pastoral farm somewhere in the Midwest—what do you see out among the barns, the fatted calves, and the oil wells? No doubt, it's a windmill. Wind power has been used by many cultures around the world for thousands of years to run water well pumps and sail ships. The industrial revolution marked the beginning of a long, slow decline for wind power but also saw the development of wind turbines for the generation of electricity.

Today the U.S. has about 1,600 MW of installed wind capacity, most of it produced from three California wind farms. As engineers improve the efficiency and affordability of wind turbines (and as interest in other-than-fossil-fuel driven generation increases), such machines are coming close to the cost of conventional utility generation.

Windmills produce kinetic energy from the air moving across the earth. As the sun heats the earth's surface and atmosphere unevenly, thermal differences and the rotation of the earth drive powerful air currents. Wind turbines convert this kinetic energy first into mechanical energy and then into electric energy by means of airfoils (blades resembling an airplane propeller), drivetrains, and of course, a generator.

The greatest advantage to wind-driven generation is that it can be sized for specific generating amounts, from a single turbine to a "wind farm" with hundreds of units. Unlike large fossil fuel plants, the land on which windmills are erected remains available for other agricultural uses. Another advantage is a complete (and obvious) lack of emissions and the fact the "fuel" is free!

The greatest drawback to these facilities is inconsistent wind speeds. (Of course, sufficient winds are not available everywhere!) Even small changes in wind speed mean large changes for power produced. Because longer airfoil blades can take advantage of variable wind speeds and so generate more power, today's wind turbines are much larger than in the past. Blades can be up to 100 feet long, to enable wind turbines to generate up to 1,300 kW or higher. The other major problem is that while wind-generated power costs a fraction more than fossil fuel-generated power, almost all expense is incurred up front, in equipment and constructions costs.

Solar

What could be a more obvious source of energy than the original source of all energy—the sun? However, as with wind-generated electricity, the prime consideration is location—the amount of useable sunlight received in any given area depends upon geographical location, time of day, season of the year, and weather (clouds). In the U.S. for instance, the desert southwest receives almost twice as much sunlight as all other areas.

Once the location is determined to be right, it's time to consider equipment. For solar energy systems to work we need solar cells—some form of collector to gather the power needed to generate electricity. Current solar energy technologies for power generation include photovoltaics and thermal systems—light and heat converters.

Photovoltaics consist of semiconductor material that directly converts sunlight into electricity. Power is produced when sunlight strikes the semiconductor material and creates an electric current. Solar thermal systems use heat to generate electricity. A system of mirrors and lenses concentrate and focus sunlight onto receivers that absorb and convert sunlight into heat;

the heat is routed to a steam generator where it's converted into electricity.

Like wind power, solar energy technologies make sense because the fuel is free and the power produced is clean, renewable, and 100% domestic. Like windmills, solar generating systems can be constructed to meet any sized requirement and are easily modified to meet changing energy needs. The drawback is that despite ongoing and impressive technological improvements, they remain more costly to build and maintain than fossil-fueled systems.

Hydroelectric power

The power derived from moving water has long been used to make work easier, and it helped make the industrial revolution possible. Today, hydropower plants range in size from single households or factories to centralized plants with capacities of 10,000 MW (Three Gorges). Electricity derived from moving water makes up about 25% of the world's total output—nearly 2.3 trillion kWh. U.S. capacity is more than 92,000 MW (10% of the country's electricity), making us the world's leading hydropower producer.

As with wind and solar, the key here is converting a natural resource—the energy in flowing water—into electricity. How much is generated depends upon how much water flows and the height difference between the plant turbines and the water's surface. The greater the flow and height (or "head"), the more electricity is produced. Most plants utilize water diverted from rivers, assisted by reservoirs to make up for seasonal fluctuations in water flow and so provide a constant electricity supply. Others use what's called pumped storage: after water flows through turbines and produces electricity, it is directed to a reservoir so that in periods of low river flows, it can be pumped into an upper reservoir and reused.

The advantages are obvious—inexpensive electricity with no pollution. What's more, water does not suffer "terminal use" in the production of electricity—it can be reused for this or for other purposes. However, hydropower plants require streams of a certain minimum flow and their reservoirs take up a lot of space.

Biomass energy

Just as solar, wind, and hydropower access the natural resources of the sun, wind, and water, so biomass energy taps energy stored in plants and other organic matter. Biomass feedstocks (normally dried, chopped vegetation, or wood wastes) can be used in power plants in ways similar to how nonrenewable fossil fuels are used to generate electricity.

Specific biomass materials include wood and wood wastes, agricultural crops and their waste byproducts, MSW, animal wastes, food processing waste, and aquatic plants, including algae. Methane—the primary component of natural gas—can be harvested from landfills. There are also so-called "energy crops"—fast-growing grasses and trees—that can be used in biomass energy conversion. Not surprisingly, wood-related industries consume most of the biomass energy, burning their wastes in furnaces and boilers to supply the energy needed to run their factories.

There are three primary techniques. In direct combustion, biomass is burned to produce steam, which turns a turbine and drives a generator to produce electricity. Gasification converts biomass into a combustible (bio)gas to drive a combined-cycle gas turbine. Pyrolysis is a system in which heated biomass is converted into pyrolysis oil and burned to generate electricity. Biomass can also be refined into fuels such as ethanol, methanol, biodiesel (as well as additives for reformulated gasoline) to run motor-driven generators.

In addition to its value as a fuel, biomass produces fewer air emissions than fossil fuels and lessens the amount of landfilled wastes; again, it's a domestically produced fuel, as well. In some areas of the world, "biomass energy crops" are grown in an effort to help make biomass energy cost competitive with fossil fuels. The major drawback to this technology is that biomass' energy potential is less concentrated than that of fossil fuels so it is not currently competitive with coal, oil, and natural gas.

Geothermal energy

If we're going to generate electricity using natural resources above and on the surface of the earth, why not dig deeply for a natural resource found beneath the surface? Therein lies geothermal energy—the heat held below the earth's crust. Brought to the surface as steam or hot water through heated, permeable rock, it can be converted into electricity. In the U.S. most geothermal resources are found out west.

Globally, there are five kinds of geothermal energy—hydrothermal fluids, hot dry rock, geopressured brines, magma, and ambient ground heat. Of them, only hydrothermal fluids have been developed commercially for power generation at this writing. How that resource is converted into electricity depends on whether the raw material is found as steam or water, and its temperature.

Hydrothermal fluids are those that are wholly or primarily steam, routed directly to a turbine to drive an electric generator. This dispenses with the need for the boilers and conventional fuels to heat water. Hydrothermal fluids above 400°F that are primarily water are sprayed into a tank held at a much lower pressure than the fluid. This causes some of the fluid to rapidly vaporize (or flash) to steam and drive a turbine and a generator. Water less than 400°F is combined with hot geothermal fluid to vaporize a secondary fluid that drives a turbine and generator.

Steam resources are obviously best, but thus far, the lone commercially developed steam field is the Geysers in northern California, which have been producing electricity since 1960. Hot water plants are more common and have become the major source of geothermal power here at home and around the world.

As with the other alternative fuels we've considered, hydrothermal power plants have minimal impact on the environment. They release few greenhouse gasses. They're quite reliable, compared to conventional power plants and in some parts of the world, are cost competitive with fossil-fueled plants.

The drawback, of course is the obvious one—if there isn't a geothermal source in your neighborhood (or your part of the world), you can't expect to be able to go out and build one!

Part 3: Electric Power Generation

Chapter 6
Prime Movers: Steam Boilers, Storage, and More

Prime Movers

The human body was the first prime mover people utilized. It was soon followed by asses, burros, dogs, elephants, oxen—and somewhere along the way, as families grew into tribes and civilizations—by other people.

People invented tools. A rock tied to a stick was not much of a screwdriver, but the Moovian Wal-Mart store was facing stiff opposition by local skin and vine merchants and King Tunk took a long time to grant permits, so hardware was scarce anyway. To make up for this, our forefathers developed the tradition of using a screwdriver for a chisel, hammer, icepick, lever and paint stirrer. (The previous owner of our home even drove his biggest screwdrivers into trees to support bird feeders.) Eventually, customers around the campfire, waiting for the new Wal-Mart to open, noticed that small nuggets of shiny material melted out of certain rocks. Teen cavers immediately saw the possibilities. They removed the rock headsets tied to their ears, borrowed their fathers' screwdrivers, and began to beat the nuggets into disks.

Figure 6-1: Cutaway Section of a Jet Engine Turbine

"Cool. These will work in those machines with gum and toys in them," grunted the inventors of the slug.

Finally, people got tired of waiting for progress to arrive and wandered off to make some of their own.

The Greek inventor, Hero of Alexandria, invented the steam turbine about 75 AD. Hero's turbine was a metal ball partially filled with water and nozzles aimed tangentially from its equator. Had he angled the nozzles lower, he probably could have invented Sputnik or begun the UFO phenomenon. Suspended on a pivot, a flame heated the turbine until the water turned to steam and thrust from the nozzles, rotating the ball.

Although Hero's reaction turbine was a simple machine, a curious toy, it was a prime mover and operated on the same principles as today's jet engines. Modern gas turbine engines are impulse turbines (Figs. 6-1, 6-2). Air is forced into the combustion chamber of the engine by the compressor vanes and fuel is injected under pressure and ignited. Forcing the expanding exhaust gases through the various "buckets" causes the shaft to turn. In an aircraft engine, the shaft turns additional fan blades to entrain outside air along with the exhaust and create thrust to fly the plane. In a gas turbine generator, the shaft connects directly to the generator. Think of it as a "gas-turbine-kebab." In typical topping-cycle cogeneration plants, the hot exhaust gases boil water to make steam for a conventional

Figure 6-2: Cutaway View Shows Turbine "Buckets"

steam turbine generator and the residual steam is used for heating or industrial processes. In a bottoming-cycle plant, the exhaust gases boil water for steam and are used first for process heating, then for electrical generation.

Today's prime movers are engines that convert potential energy—fuel—into kinetic energy—motion. In approximate historical order of development, artificial prime movers include water wheels and turbines, steam turbines and reciprocating steam engines with their fireboxes and boilers, internal combustion gasoline and diesel engines, nuclear steam turbines, and gas turbines. While the steam turbine was invented almost 1,700 years before the first practical steam engine, jet engines and nuclear reactors, developed in secret during World War II took about %1 as long to become practical. The jet engine-on-a-stick approach to power generation was first applied to railroad locomotives about the same time the first post-war jets flew.

Today's prime movers for electric power generation include fossil-fired steam turbines, natural gas turbines, nuclear reactor steam turbines, and water turbines. Diesel engine systems are even utilized. (Fig. 6-3) Clearly, some kind of turbine is the prime mover that turns the generator to pro-

Figure 6-3: Diesel Engine System

duce almost all electricity. Tidal generation is a special case of hydroelectric "turbine."

With renewable energy resources other than hydroelectric, there may not be a clear cut prime mover. For example, in geothermal and solar, the energy source is already in a kinetic form—steam and electromagnetic radiation—so there is no transformation from potential energy to energy of motion that defines the term "prime mover." Solar generation takes two forms—thermal and photovoltaic. In thermal solar, reflectors focus sunlight on pipes designed to boil water for—you guessed it—a steam turbine. Photovoltaic solar uses silicon photocells arrayed as solar batteries to generate electricity directly.

Steam boilers

Boilers are closed vessels in which water is heated and steam is generated to drive a turbine. Boilers consist of a boiler shell (the body of the boiler), laced with tubes in which water is circulated as it is turned into steam.

In fossil fuel-fired steam turbine generators, fuel is burned in a furnace beneath the boiler. This is where fuel's chemical energy is converted into heat energy. The type and grade of fuel used determines the highly specialized furnace designs. Boiler design follows the furnace arrangement.

Boilers are designed to absorb all possible heat coming from the fuel and their efficiency is measured by the percentage of heat absorbed. If all of it were absorbed, the boiler would be rated 100% efficient.

There are two types of boilers. In fire-tube boilers, products of combustion are circulated through tubes that are surrounded by water. This is a low-pressure type of boiler, furnishing hot water at pressures not exceeding 160 pounds per square inch (psi) or at temperatures not exceeding 250 degrees F (121 degrees C) or steam at pressures not exceeding 15 psi. Its applications are limited. In a water tube boiler, water is circulated through the tubes and the gases of combustion surround the tubes. These can be either low or high pressure, furnishing steam at pressure in excess of 15 psi or hot water at temperatures in excess of 250 degrees F (121 degrees C) or at pressures in excess of 160 psi. Low-pressure boilers are preferred because their small amounts of circulated water can be quickly heated and circulated. Water tube boilers are further grouped into straight-tube boilers and bent-tube boilers.

Storage and more

Storing electricity—what could be easier? There's the flashlight in the kitchen drawer and in the glove box of the car—(oh yeah, that one's dead)—and of course the battery under the hood. In truth, storing electricity as electricity is a very iffy proposition. Batteries actually store electricity as potential chemical energy. Only capacitors store an actual charge and chokes—coils that act as an electrical flywheel—can store current. High-temperature superconductors, moving from the headlines to practical applications, offer some promise for storing electrical energy in special cases.

At the present time, energy suitable for generating electricity is best stored in its potential chemical or nuclear form. Pumped storage facilities are a distant, but vital second. (Besides, generators run just as efficiently as motors and water turbines operate nearly as well as pumps.) Pumped storage stations usually must be purpose-built as in most cases, the water below a hydroelectric dam disappears down river and is unavailable to be pumped back up to the reservoir. The Northfield Mountain Pumped

Storage Station described in chapter 13 has its "runners" about 60 feet below the Connecticut river from which it draws and discharges, assuring an adequate supply of water nearly year round.

With the recent shift in focus from storage batteries to fuel cells, the future of the electric car may have been revived. An ideal fuel cell reacts oxygen and hydrogen to yield water and electricity or water and electricity may be turned back into oxygen and hydrogen. As currently envisioned, fairly conventional fuels such as alcohols will react with atmospheric oxygen to generate electricity—the latest incarnation of a prime mover! An electric car with lead-acid ("HPb") storage batteries is literally a "lead sled." The potentially satisfying performance of the electric motor is sapped by having to drag around the equivalent of six passengers with only the driver on board (and it is a disaster waiting to happen if it is in a serious accident). The downside of the fuel cell electric car is that using conventional fuels and atmospheric oxygen will generate emissions similar to the internal combustion engine.

Although hydrogen got a bad rap in the Hindenburg dirigible disaster, it is the ideal fuel for fuel cells. It could be generated at home from water and photovoltaic batteries any sunny day and supplemented by the ac mains.

Once they become practical, high-temperature superconducting transmission lines will be the next driving force for technological change in the electric power industry, allowing coal plants to be sited adjacent to the coal mines and nuclear plants to be built on seismically stable ground away from population centers. Transmission energy losses (estimated to be at least 5%) would be eliminated, providing a significant boost in system efficiency.

Chapter 7
Furnaces and Boilers

In fossil fuel-fired steam turbine generators, a furnace is where the fuel is burned and its chemical energy is converted into heat energy. The boiler is where the steam is generated. The type and grade of fuel used determines the highly specialized furnace designs. Boiler design follows the furnace arrangement.

In a perfect world, solid fuel—coal—would burn according to the following equation:

Eq. 7.1 $C\ (s) + {}^{2}O\ (g) \Rightarrow CO_2\ (g) + \text{heat}$

where:

$C\ (s)$ is carbon as pulverized coal

${}^{2}O\ (g)$ is oxygen gas

$CO_2\ (g)$ is carbon dioxide gas

Natural gas would burn as follows:

Eq. 7.2 $CH_4\ (g) + 2O_2\ (g) \Rightarrow CO_2\ (g) + 2H_2O\ (g) + \text{heat}$

where:
CH_4 (g) is natural gas (methane)
H_2O (g) water is water vapor

Some emergency back-up generators run on propane (bottle gas):

Eq. 7.3 C_3H_8 (g) + $5O_2$ (g) => $3CO_2$ (g) + $4H_2O$ (g) + heat
where:
C_3H_8 (g) is propane gas

Methanol (wood alcohol) and ethanol (grain alcohol) have been introduced into gasolines as additives and are used as fuels in their own right:

Eq. 7.4 $4CH_3OH$ (l) + $6O_2$ (g) => $4CO_2$ (g) + $8H_2O$ (g) + heat
where:
CH_3OH (l) is liquid methyl alcohol

Eq. 7.5 C_2H_5OH (l) + $3O_2$ (g) => $2CO_2$ (g) + $3H_2O$ (g) + heat
where:
C_2H_5OH (l) is liquid ethyl alcohol

Finally, thanks to a Nazi dirigible and a radio reporter who lost his detachment in a crisis, there is one more fuel that is rarely mentioned: H_2.

Eq. 7.6 H_2 (g) + O_2 (g) => $2H_20$ (g) + heat
where:
H_2 (g) is hydrogen gas

Furnaces

Furnaces are designed and operated to bring combustible gases into contact with the correct amount of air to establish and maintain a temperature higher than the fuel's ignition temperature. How well this is done determines combustion efficiency. It's the same here as it is for any fire:

Figure 7-1: Close-up of 9 MW Steam Turbine Burner

what's needed are fuel, oxygen, and heat, and each of them in the correct mixture. Fuel should be completely burned, allowing little or none of it to escape unburned and keeping excess air to a minimum, in order to deliver heat to the boiler with as little temperature loss as possible.

Furnace operating conditions and design must meet the particular conditions of each application and allow for type and grade of fuel. External heat is used for ignition. Thereafter, heat for the continuing chemical reaction is derived from the fuel as well as the walls of the furnace. If furnace temperatures drop below the ignition temperature (and that varies, depending on the fuel), then combustion will become inefficient or stop. If, as a result, gases in the boiler tubes are cooled by inefficient furnace combustion, fuel is wasted and soot builds up in the boiler tubes—a messy and expensive maintenance job!

Furnace walls are constructed of fire-resistant bricks and assembled with bonding methods able to withstand extreme heat. How extreme? Furnace temperatures can reach 3,000°F. Temperature variance is another stress issue. As furnaces go from cold starts to operating temperatures (and back again, over and over), cracks or furnace wall failure can occur. A lining made of high-grade tile or brick—sometimes with a refractory layer applied—can help control the effects of high temperature, and the effects

Figure 7-2: Steam Turbine Furnace

of abrasive fuels or combustion gas constituents. Cooling systems (such as water-cooled walls) are often added to offset massive temperature changes.

Furnaces are modified to accommodate different fuels. Coal requires a *hopper bottom furnace*—a furnace bottom with inclined sides that form a collection area (a hopper) to make the collection and removal of ash much easier—and an elevated grate on which coal is burned—the *stoker*—to allow maximum fuel-air balance and help facilitate ash removal and additions of fresh coal. When utilizing oil or gas (or pulverized coal), fuel is injected through burner jets. They mix fuel with air in correct proportions then shoot the mixture into the furnace. The air supply—called the *draft*—varies according to the type of fuel used and has to be provided, either through design of the equipment (*natural draft*) or by mechanical *induced-draft* or *forced-draft* equipment. As has been said: "induced draft sucks and forced draft blows." Air initially mixed with the fuel is called the *primary air*; air added in the furnace is called *secondary air* (Figs. 7-1, 7-2, 7-3).

Boilers

Boilers—steam generators—are closed vessels in which water is heated by the application of heat from the furnace, and steam is generated

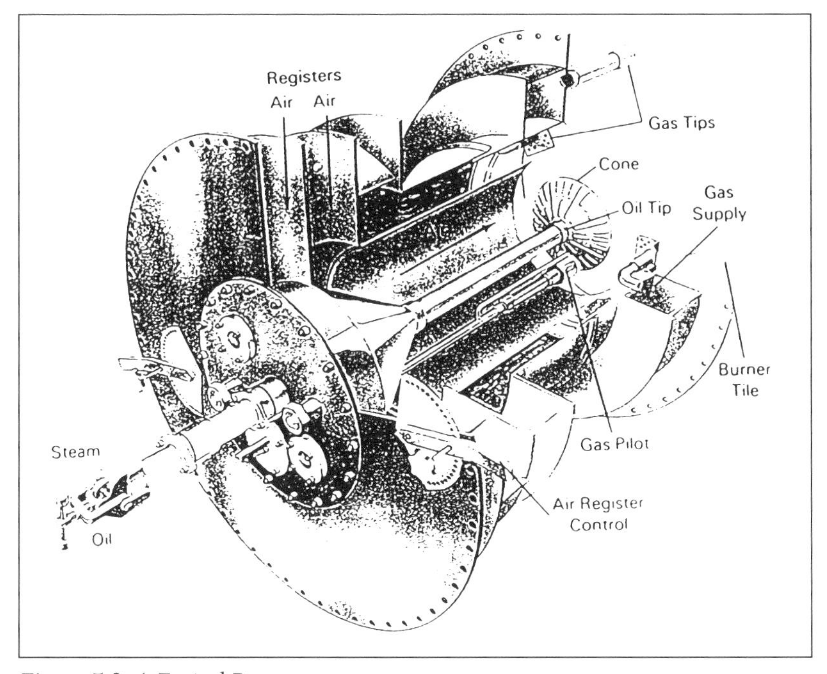

Figure 7-3: A Typical Burner

(and sometimes superheated, under pressure or vacuum). Boilers consist of a boiler shell (the body of the boiler), which is laced with tubes in which water is circulated as it's turned into steam.

Boilers are designed to absorb all possible heat coming from the fuel and their efficiency is measured by the percentage of heat absorbed. If all of it were absorbed, the boiler would be rated 100% efficient.

There are two types of boilers. In *fire-tube boilers*, products of combustion are circulated through tubes that are surrounded by water. This is a low pressure type of boiler furnishing hot water at pressures not exceeding 160 psi or at temperatures not exceeding 250°F (121°C) or steam at pressures not exceeding 15 psi. Its applications are limited. In a *water tube boiler,* water is circulated through the tubes and the gases of combustion surround the tubes. These can be either low pressure or high pressure—furnishing steam at pressure in excess of 15 psi or hot water at temperatures in excess of 250°F (121°C) or at pressures in excess of 160 psi. Low-pressure boilers are preferred because their small amounts of circulated water can be quickly heated and circulated. Water tube boilers are further

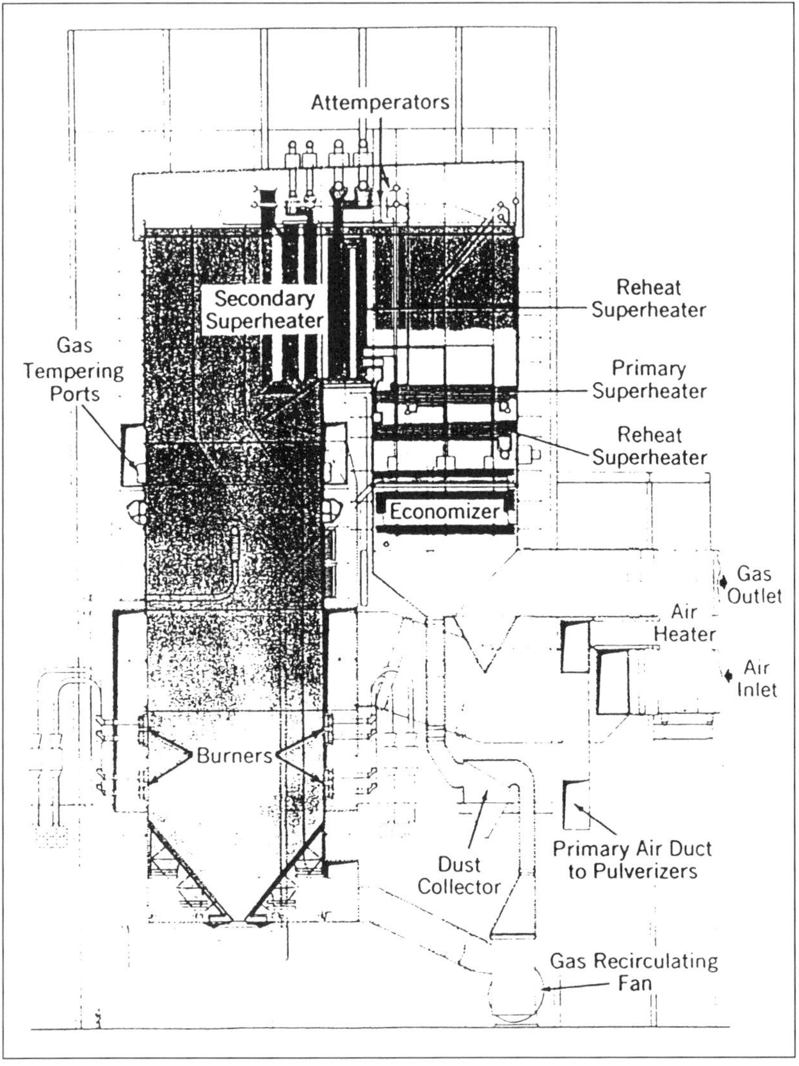

Figure 7-4: A Typical Boiler

grouped into *straight-tube boilers* and *bent-tube boilers* (See Fig. 7-4).

In boilers used in electric plants, the heating surface consists of two drums connected by circulating tubes. A *water drum* at the lower part of the boiler is connected by seamless steel tubes to a *steam drum* located diagonally at the upper part of the boiler. Circulating tubes—smaller than those from the water drum—are gathered in groups so that the water circulated through the tubes absorbs heat from the furnace. As the rate of circulation increases, the output of the boiler increases.

In a teakettle, heat brings the water to a boil, producing steam. More heat produces more steam. Standard boiling temperature of water is 212°F. If external air pressure is placed on the water, however, the water must be hotter than 212°F to attain a boil. Conversely, if external pressure is *lowered* from standard air pressure, then water will boil at a lower temperature.

When a liquid is heated to a boil, its physical state changes—it turns to vapor, or steam. It takes 970 Btu for one pound of water at the boiling point to change into steam of the same temperature. The 970 Btu threshold is known as the *heat of vaporization.*

Regular steam—the kind coming out the spout of a teakettle—is called *saturated steam.* When a container holding water and steam continues to be heated after all the water has turned to steam, the steam will become *superheated.* Superheated steam behaves like a gas and is in a far more stable condition than regular steam. A large amount of heat can be lost from superheated steam before condensation or liquefaction occurs. Regular saturated steam generally loses a small amount of heat before condensing back into water.

In a boiler, saturated steam passes from the steam drum to a separate set of tubes inside the boiler to receive more heat and become superheated. The superheated steam is then collected in another drum, from which it passes on to the turbine room. In the superheater, the temperature of the steam is increased but the pressure remains the same or drops slightly due to friction in the superheater tubes and piping to the turbine room.

Water used in boilers—*feedwater*—must be as pure as possible. Just as hard water coming from the tap in a residence can cause scale to form on the bathroom fixtures and clog up the showerhead, water with impurities in a high-pressure boiler can cause scale and a buildup of impurities that can lead to inefficient boiler function and to maintenance problems.

Chemically pure water is extremely rare, however, because water is an almost perfect solvent—almost all substances are soluble in it. Although a variety of mineral salts or acids may be present in natural water, only a few are found in sufficient quantities to pose a problem in boiler feedwater. These include salts of calcium, magnesium, sodium, and a few other materials that can cause scale deposits and sludge to accumulate in the boiler's metal tubes and piping as well as corrosion of the metal itself. Mud, clay, sewage, and other waste products can be present in water as suspended solids and can cause scale formation. Oil likewise can cause corrosion, deposits, or foaming in the boiler. Gases such as CO_2 can accelerate corrosion of metal parts.

There are several ways to combat these impurities in water. Settling tanks with filtration and water-flushing devices can help remove sediment such as mud or sand. Settling tanks with chemical coagulants may be used to turn sediment in the water into a jelly-like substance that can be filtered out. Evaporation or distillation can remove all forms of impurities. Finally, chemical treatment can remove impurities such as calcium or magnesium salts.

Treatment of feedwater can be expensive, but it is more economical than the reduction in boiler efficiency, maintenance problems, and down time that comes from using "raw" water as feedwater. Just as a household clothes iron will offer a longer period of worry-free use when it is filled with distilled water rather than tapwater, a utility boiler will run for a longer time without maintenance problems when it is filled with clean water.

Chapter 8
Steam Turbines

As we have seen, prime movers convert potential energy into kinetic energy to drive generators that "make" electricity. In most modern power plants, the potential energy is heat energy of the furnace/boiler and the kinetic energy is the mechanical energy of a steam turbine (Fig 8-1).

Steam turbines are the heart of any large, centralized power plant. They have dominated the U.S. power industry from its beginnings and will continue to do so after deregulation is in place. They have helped to electrify America because they can be constructed with capacities far larger than is possible, or even practical, with the other prime movers.

Yet, in both theory and application, steam turbines work in ways similar to a basic water wheel that is turned by a stream of flowing water. Water falling across the paddles of the wheel causes it to turn and "do work." Steam produced by the boiler pushes across the turbine blades, turning the turbine and spinning the generator. The higher the turning speed and the larger the turbine, the more power it produces. One advantage that steam turbines possess over even a modern waterwheel is that they can achieve greater efficiencies because their size and power is not limited by the pressure "head" above the wheel. Modern alloys and tech-

Figure 8-1: Steam Turbine

nology culminating in more than 100 years of experience permit the use of pressures, temperatures, and velocities unimaginable in a hydroelectric plant.

As noted in chapter 5, a variety of fuels can power electric generating facilities (coal being the most common). As fuel is burned, it brings water to boil in a boiler, creating steam. Steam is directed into the turbine through jets, under pressure, against the turbine blades. This causes rotation of the turbine shaft (or rotor) to which the blades are attached. Greater flows of steam produce greater forces exerted on the turbine blades, resulting in greater speeds and more work performed. Returning to our example of the water wheel turbine, imagine that a water source can be maintained not only at a constant rate, but also increased at will!

Steam turbine efficiency remains at about 40% despite improvements in design and controls. Throughout time, for instance, turbine blades have been modified for greater efficiencies. Steam bottled up in the boiler is like water in a reservoir. In order to release its potential energy it must thrust against the turbine blades, expanding as it does (Fig. 8-2). It was quickly discovered that after steam pushed the blades and turned the turbine shaft, it still contained useable energy. Unused steam energy escaping into the

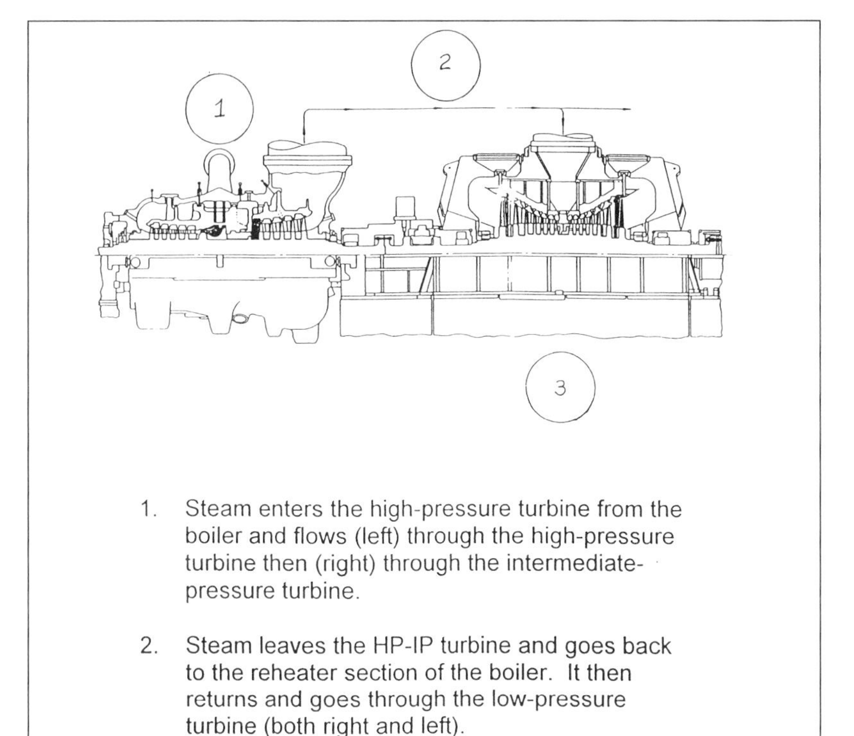

1. Steam enters the high-pressure turbine from the boiler and flows (left) through the high-pressure turbine then (right) through the intermediate-pressure turbine.
2. Steam leaves the HP-IP turbine and goes back to the reheater section of the boiler. It then returns and goes through the low-pressure turbine (both right and left).
3. Steam exhausts to the condenser from the low-pressure turbine

Figure 8-2: Steam Flow Through a Turbine

atmosphere represented a lost resource. Curved turbine blades were more effective in harnessing steam flow and conducive to directing steam to additional sets of blades, thus more completely using the steam energy.

Modern steam turbines carry several sets of curved blades in varying sizes to make the best use of the energy provided by the expanding steam. Such blade sets are called stages. Steam turbines are rated by the number of available stages. That number depends upon the ability of the rotor to create kinetic energy as closely equal to the potential energy of the available steam as practical. This energy-converting capacity is designed into turbines by rotor length and diameter, revolutions per minute (rpm), and the number and design of the blades and their stages.

Internal corrosion is not a major issue for steam turbines despite their exposure to super-heated steam. The preheated liquid water in the boiler is very pure and at these temperatures has little capacity to carry dissolved O_2. As air has little force to turn the turbine, special valves bleed it out of the system wherever it may accumulate. Nickel or chromium stainless steel turbine blades and rotor and blade stage casings resist corrosion. Such materials are high strength, heat resistant, with low coefficients of expansion.

Like any gas, steam contracts as it cools. Temperature variations from inlet to output cause blades, casings, and rotor to expand and contract unequally every inch of the way. Distortions could occur given the different temperatures at the inlet and exhaust ends of the turbine and must be allowed for in the design and operation of the steam turbine. Main bearings are lubricated with oil under pressure and a thrust bearing at the high-pressure end guards against uncompensated axial thrust.

Just as turbine blades are arranged in stages, so the stages are arrayed in several sections. Each section contains blades of different sizes. This, too, is done to achieve maximum efficiency—as the steam moves through each section it loses pressure and potential energy. To compensate for this, blade length is increased so that each section of the turbine produces the same amount of kinetic energy (or as close as possible) as the section before it. This also facilitates smoother turbine rotation and reduces stress on the turbine rotor shaft that would occur if energy conversion in each section were significantly different.

The first section—the high-pressure section—is where the high-intensity steam first enters the turbine, and so this section contains the smallest blades. Steam moves to the intermediate-pressure section, which contains larger blades, and then into the low-pressure section, which contains the largest blades. All stages/sections are attached to a single rotor. Sections are measured in pounds per square inch (psi) of steam. For instance, a turbine with a total of 20 stages may first receive steam at the high-pressure end—400 psi; steam at the 10th stage—80 psi; at the 15th stage—about 30 psi; and at the 20th—perhaps 5 psi.

The steel shaft of the turbine rotor is tough and quite heavy so it can hold its own weight as well as the weight of the blades and endure some amount of contraction and expansion. For this reason, also, the rotor sags somewhat when the turbine is stopped and cooled. Turbines must be started slowly, with heat introduced gradually to permit the shaft to heat up (and straighten up) as it turns. This may take several hours (and explains why electric plants are slow to "come up" after a trip or maintenance shutdown). Slowing and stopping a turbine requires the same routine in reverse. Once stopped, time must be allowed for the heat within the turbine to dissipate. Sudden starts or stops—trips—can damage or ruin a turbine.

Turbines are further distinguished by how they distribute steam along their rotor sections. Large turbines designed to "shoot" the steam in a direction approximately parallel to the shaft are axial-flow turbines. Those that flow the steam approximately tangential to the rim of the rotor are called tangential-flow turbines. Smaller turbines that flow steam in a radial manner—inward, toward the shaft—are called radial-flow turbines.

Turbines in which nearly all the steam flows in the same direction are single-flow turbines. When steam flows are divided and directed parallel to the rotor in opposite directions the turbine is known as a double-flow turbine. Double-flow turbines are better able to turn large generators when a single-flow turbine would be impracticably large or too expensive.

Large generation requirements during a long period of time require the use of a topping turbine. This turbine operates at the highest possible pressure (1,000-2,000 psi), receiving the entire volume of steam and driving the generator. The exhaust steam (perhaps 500-600 psi) is strong enough to drive additional single-flow or double-flow turbines that operate a generator smaller than the one driven by the topping turbine.

Turbines can increase the overall efficiency of a power plant by "donating" steam ahead of their exhaust to other equipment, or heating water for other plant uses. This practice is called bleeding or the turbine regenerative cycle. How much steam is extracted, where, and at what pressure, varies depending on the need.

Condensers

At the exhaust end of a turbine is the condenser (Fig. 8-3). Where the exhaust steam is not used for heating or industrial processes, condensers convert waste steam back into water as it exits through the last section of the turbine. Condensing the steam creates a vacuum, increasing the pres-

Flow through a Condenser

1. Steam enters the condenser from the low-pressure turbine.

2. Condensate (condensed steam) leaves the condenser and is collected in the hot well, from where it is pumped back to the boiler by the boiler feed pump.

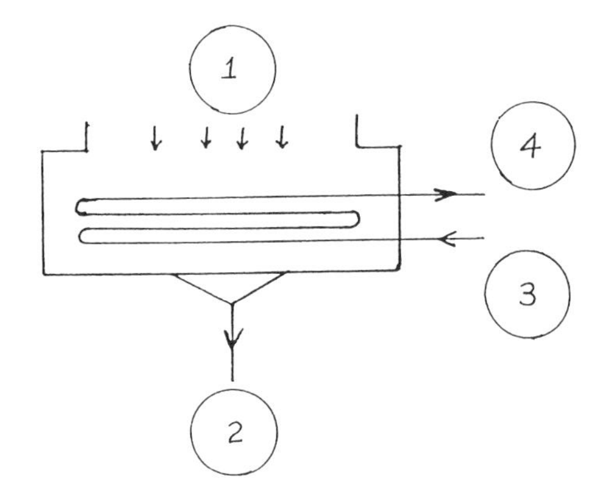

3. Cool circulating water enters from the bottom (basin) of the cooling tower.

4. Hot circulating water is pumped to the top of the cooling tower.

Figure 8-3: Flow Through a Condenser

sure differential between the input and output of the turbine, which in turn raises efficiency. Steam is the gaseous form of H_2O. Turning steam back into water releases exactly as much heat as it took to turn it into a gas. Heat exchangers in the condensers preheat water on its way to the boiler. While steam turbine plants practically cannot exceed 40% efficiency, every technique must be used to its full advantage or efficiency erodes quickly.

The two main types of condensers work in completely different ways toward different ends. With heat, energy flow is always from higher temperature to lower temperature. In a jet condenser, exhaust steam directly contacts the cooling water. Exhaust steam flows through a fine water spray. In a surface condenser, exhaust steam and cooling water do not come into direct contact with one another. Instead, metal tubes contacted by the steam conduct heat to the liquid water flowing on the other side. The hot steam runs along one side of the walls warming the cool water on the other side. In both cases, water recovered from the steam is still very hot. In a jet condenser, the residual steam and water vapor is wasted to the atmosphere. In a surface condenser, the condensed steam returns to the boiler and the heated cooling water makes up for any lost steam.

In surface condenser systems, the hot water is reused as boiler feedwater. It takes far less fuel and heat energy to convert hot water into steam than it would take to boil cool water. This is not possible in the case of jet condensers where steam condensate mixes with the cooling water. Boilers in such systems must use fresh feedwater, increasing both water and fuel costs. Therefore, while surface condensate systems are more expensive to install and take up more room, they save on fuel and water costs.

Chapter 9
Generators

Large or small, a generator is a magnet spinning inside a coil of wire, producing an electrical charge that is captured in the wires. Magnetism—more exactly, *electro-magnetism*—is the essential concept here. Unlike "natural" or permanent magnets, power plant generators make use of magnetism that is artificially created and disappears when the "creating force" (provided by the turbine) is removed.

Electricity flowing through a wire creates a magnetic field. Because this magnetic field is weak, a generator's wires are tightly wound and stacked atop one another in *rings*. (They are often called *windings*, *coils*, or *turns*). Such an arrangement concentrates the *magnetic lines of force* in the rings and so increases the magnetic effect. A core of soft iron is used to noticeably increase it. The magnetic lines of force encircle the ring much as your fingers encircle a wire when you grasp it. This device, then, an electromagnet. Turn off the current and the magnetism disappears. Turn it on, and lines of force enter the electromagnet at its South Pole and exit at its North Pole.

The relationship between electricity and magnetism—that is, producing the former with the help of the latter—is not the only relationship they share.

When a conductor (a wire, let us say) is moved through a magnetic field, it produces an electrical pressure (*induced voltage*) in the conductor. The magnetic field acts as a force resisting the movement of the conductor. It requires work to do so. The energy needed to push the wire through the magnetic field is equal to the electrical energy generated in the conductor minus the energy lost in the conversion. In this way, *mechanical work* is converted into electricity.

As more powerful electromagnets are employed, greater voltage (electrical pressure) is induced into the conductor, as more mechanical work is needed to move the conductor through the (stronger) magnetic field. Lengthening the conductor produces more voltage because a longer conductor cuts through more lines of force. If the conductor is moved through the magnetic field faster, that, too, increases voltage because it requires more work.

Definitions of Electrical Quantities: Ohms on the Range

Before applying this knowledge to the workings of a generator in an electric power plant, it is best that we revisit a few definitions.

Electric pressure is expressed in Vs or kVs (similar to water pressure expressed in psi). Electrical current is expressed in amperes (similar to water current expressed in gallons per minute). Electrical resistance is expressed in Ohms (similar to the friction encountered by the flow of water in a pipe). In electric circuits, resistance has a similar loss of electrical pressure expressed in Vs:

Pressure $E = I \times R$ = Current (amperes) x Resistance (ohms)
Where:

E = EMF in Volts
I = Current in Amps
R = resistance in Ohms

Ohm's Law, again, expresses the relationship that exists among these electrical quantities (the flow of the current varies directly with the pressure and inversely to the resistance):

$$\text{Current (amperes)} = \frac{\text{Pressure (Vs)}}{\text{Resistance (Ohms)}} \qquad I = \frac{E}{R}$$

Circuits

Electrical circuits are the paths over which electrical current flows. They originate at the positive terminal of the supply source, flow through the conducting wires and the device or devices using the energy (often called *appliances*), through more conducting wires to the second (negative) terminal of the source, and back through the source to the beginning of the circuit. The two basic types of circuits are the *series circuit* and the *multiple or parallel circuit.* Other circuits are combinations or variations of these. This information about circuits applies to devices that produce electrical energy—(generators in this case, though it could refer to batteries)—and devices that receive and use electrical energy (appliances).

In a *series circuit*, all elements are connected in succession. The current that passes through one of the parts passes through all of the parts. All appliances on the series circuit successively receive their energy from the same source, each using a part of the electrical pressure coming out of the source according to its need. This means that the total energy all the appliances need to receive from the source must add up to the original voltage supplied by the source. This is complicated by the fact that as current flows through a circuit, it loses electrical pressure from one end of the circuit to the other due to resistance. In addition, electrical pressure drops when current flows through an appliance—usually a drop much larger than is seen when electricity passes through wires because the resistance in an appliance is much greater.

In a *multiple* (or *parallel*) circuit, all components receive the full line voltage. Current flowing through each part of the circuit is impacted by resistance only in its part of the circuit. In addition, with multiple paths for the current to follow, resistance is reduced and current flow is increased, raising the conducting ability of the circuit.

Series and parallel circuits differ in their wiring, then, and parallel circuits are more efficient (Figs. 9-1, 9-2, 9-3).

If we know the amount of current flowing through a circuit and we know its resistance, then we can determine the power necessary to overcome that resistance. To the extent that we *overpower* a circuit—where electrical energy is not converted to mechanical work of some kind—it is converted to heat energy and lost. Heat, related to the amount of electri-

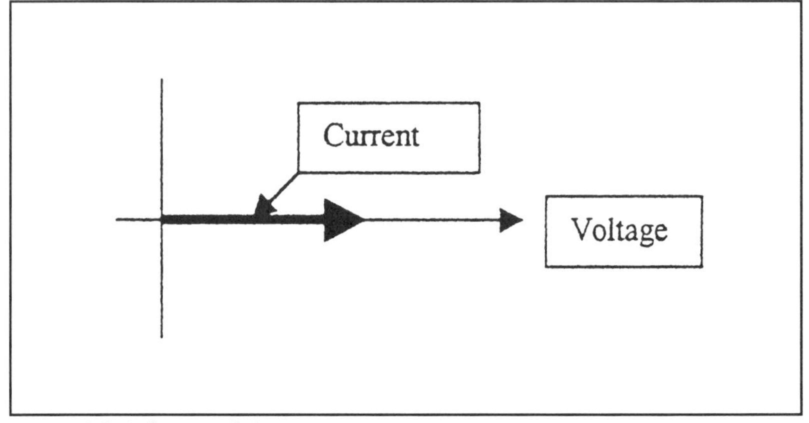

Figure 9-1: Voltage and Current

cal resistance encountered, is likened to the heat developed by friction, which also represents a loss. An example of this situation is the heat lost from the wires carrying current from a generator to an electric motor. (In determining this, the resistance of the motor wires must be included because there are losses in these wires just as there are in any wires). If resistance remains constant and wire current is doubled, heat loss is quadrupled.

Alternating Current Generators

As has been noted, *direct current* (dc) circuits are so named because electrons flow in one direction. Batteries are an example of this. Modern electric power plants use *alternating current* (ac)—with electrons flowing back and forth—because this permits transmission of electricity over much greater distances. Dc transmissions are useful for distances of a mile or less. Today's ac technology provides electricity by a *polyphase* ac generator, with three phases for smooth power transfer. Electricity is generated and sent to substations to regulate and deliver the type of power required.

Producing voltage by moving a conductor up and down through a magnetic field is not practical for even small-scale production of electricity. What is needed to light up a city is voltage—electrical pressure—in a strong, continuous stream. This is achieved by mounting the conductor between insulated discs, which are rotated by an external machine in the magnetic field.

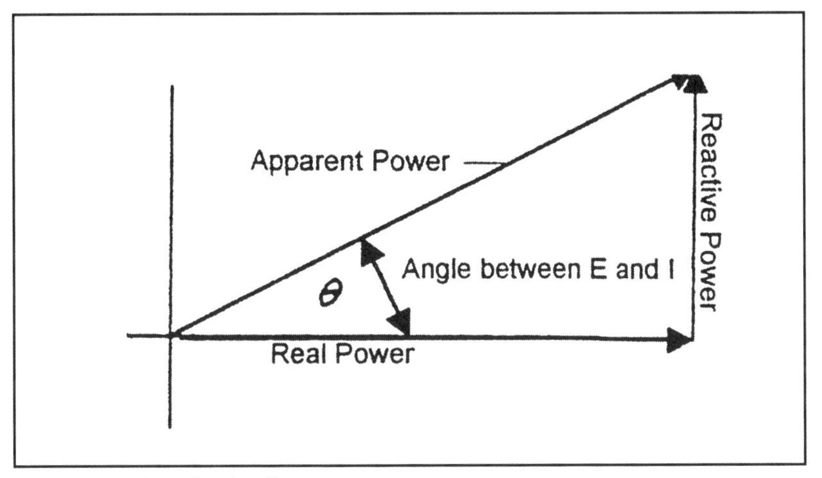

Figure 9-2: Three kinds of power

In a *single-phase generator*, a single conductor rotates at a uniform speed, producing a cycle of voltages. The voltage is zero when the conductor moves parallel to the magnetic field, then rises as it passes through the field, first one way and then the other. The current in the circuit rises and falls with the voltage flowing alternately one way and the other producing ac.

If a second conductor is added, it will produce voltage (and current) identical to the first, thereby doubling the output (and the required work) of the generator. These are considered conductors *in series*. They may be independently connected to outside circuits or connected to a single circuit. Some are connected in parallel (for greater current output) or in series (for greater voltage). However, the amount of mechanical work changed to electrical energy remains the same.

Voltage is rated according to the number of cycles completed per second (*i.e.*, 60 cycles per second). Voltage of 60 cycles per second will cause a 60-cycle current to flow. The number of cycles per second is the current's frequency.

Two-phase generators configure two single-phase generators whose armatures are mounted on the rotor shaft, rotating through the same magnetic field, always at right angles to one another. Phase 2 is always a quarter cycle behind phase one due to the relative mechanical positions of the

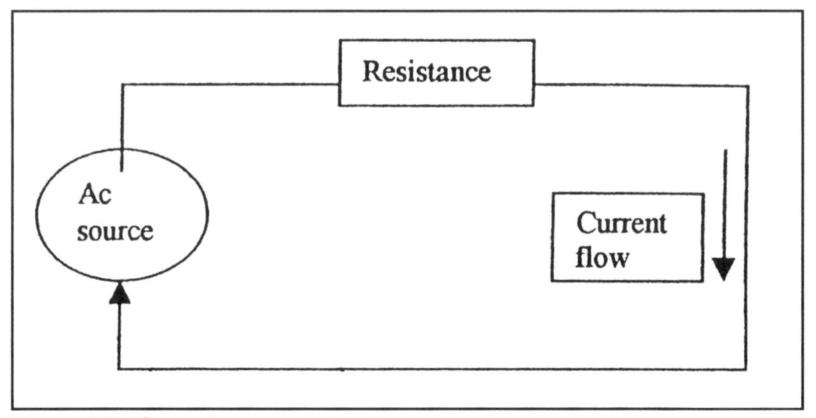

Figure 9-3: Alternating current circuits

armatures. *Three-phase generators* are also common but there is little advantage to configuring generators with more than three phases, with the exception of six-phase machines used to convert ac to dc.

Internal combustion or steam engines drive small generators (those of 1,000 kW or less). Large ac generators (1,000 kW or more) driven by steam turbines depend upon the mass of the rotor to maintain momentum needed to produce smooth, continuous voltage. Increasing generator output means increasing core length (within limits of safety, material strength, and rotation speeds). A typical example:

In an 1,800 rpm, horizontal-shaft generator with a rotor core diameter of six feet, length of 20 feet, and a turning speed (at the rotor periphery) of 35,000 feet per minute, material at the periphery experiences a centrifugal force of 2,000 pounds!

As with any machine, energy input and output differ due to losses from various sources. These include friction (mechanical) from rotating parts, as well as the interaction between the rotor and air in the gap between the rotor and the *stator* (the stationary portion of the magnetic circuit). Electrical losses are measured in the steel core of the stator and are caused by the alternating magnetic field's effect on its molecules and on the current flowing in the conductors. All losses are experienced as heat. Generators are usually cooled by fans mounted on the rotor, blowing air over the coils. In some machines, air is replaced with H_2 or helium gas for more effective cooling.

Safety Considerations

Because the generation of electric power involves big machines and life-ending amounts of electricity, safety precautions are incorporated in the design and construction of the major components (boilers, turbines, generators) and their attendant systems. In addition, precautions are available to limit damage and injury (as much as possible) in the event something untoward happens.

Circuit breakers operate automatically to de-energize components when overload, high temperature, or other "faults" occur. The physical separation of components and isolation from other systems helps to limit potential damage, should one of them develop problems. Even generators are separated from one another in plants with more than one.

Where a generator consists of three separate phases (circuits), the common connection is normally grounded and each phase is separated from the others by floors, levels, or some fire- and explosion-proof barrier. Circuit breakers and transformers are likewise isolated.

Further protection may come from chemical spray systems that are activated by automatic temperature and "over-current" devices.

Grounding — literally, connection to the earth — is the best ongoing protection for man and machine. As long as all components are anchored into the dirt, a first line of protection is in place. It also reduces strains placed on insulation and provides a discharge path for lightning arrestors.

Chapter 10
Fossil-fired Steam Plant

By the dawn of the Twentieth Century, it became as clear as incandescent light that electric power was here to stay. Consequently, small generating companies began to interconnect and evolve into the regional utilities and cooperatives we still refer to as "the electric companies." Technology evolved as well. Generators grew from hundreds to thousands of Watts. Crude hydropower arrays and small steam engines erected close to end users gave way to massive, centrally located base load units—so named because they enabled production of a continuous stream of electricity at full capacity nearly all the time. Today, fossil-fired steam power plants remain the most efficient and economical choice for the large capacity plants that continue to serve the majority of electric customers in the U.S.

Steam plants offer flexibility that generating companies and utilities need. Various types of fuel can be utilized. Boilers and furnaces are designed and constructed according to the fuel selected. Fuel choice also impacts operating and maintenance costs and procedures, as well as the costs of purchasing and transporting the fuel to the power plant and disposing of waste products. Increasing local, state, and federal regulation of emissions has to be added to the cost equation.

Because construction designs have become standardized and technolo-

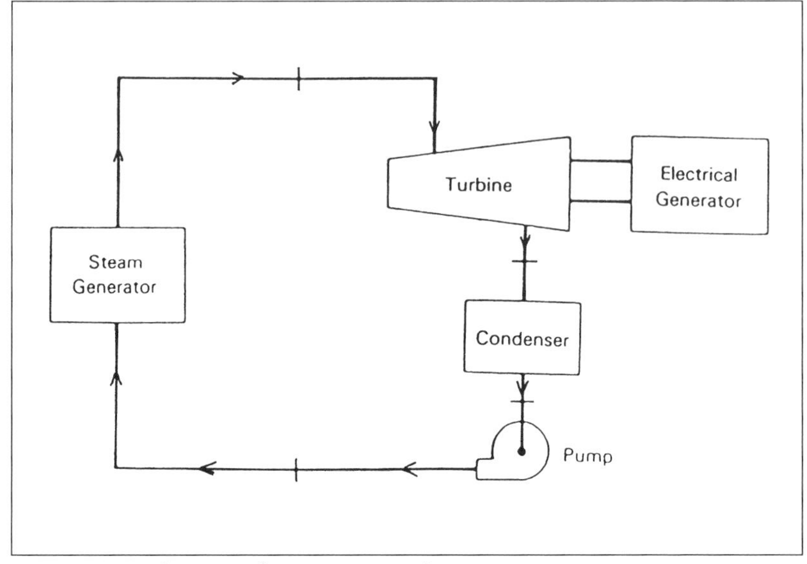

Figure 10-1: Schematic of a Basic Power Plant

gy has been proven, function can easily follow need. Does the desired load require one boiler, one turbine, and one generator? Several boilers feeding a common "steam header" to supply several turbines and generators? Such "unit-type construction" allows even the largest power plant to be tailored to serve today's needs and to be adjusted for what may be required tomorrow (Figs. 10-1, 10-2).

Construction costs vary depending upon location and what exactly you are erecting. Most power plants are built in remote areas where materials have to be brought in from great distances. Specialty labor may have to be imported. Nuclear plants have been built to greater specifications and include redundant systems. This increases their cost. The largest ongoing cost is fuel—for any power plant. Fuel considerations—transportation, storage, and handling—have to include future availability, and both current future possible environmental restrictions, and not just price. Even the financing used to build a plant is a cost consideration. Once it is erected, operations and maintenance (fixed costs) must be reviewed on an ongoing basis.

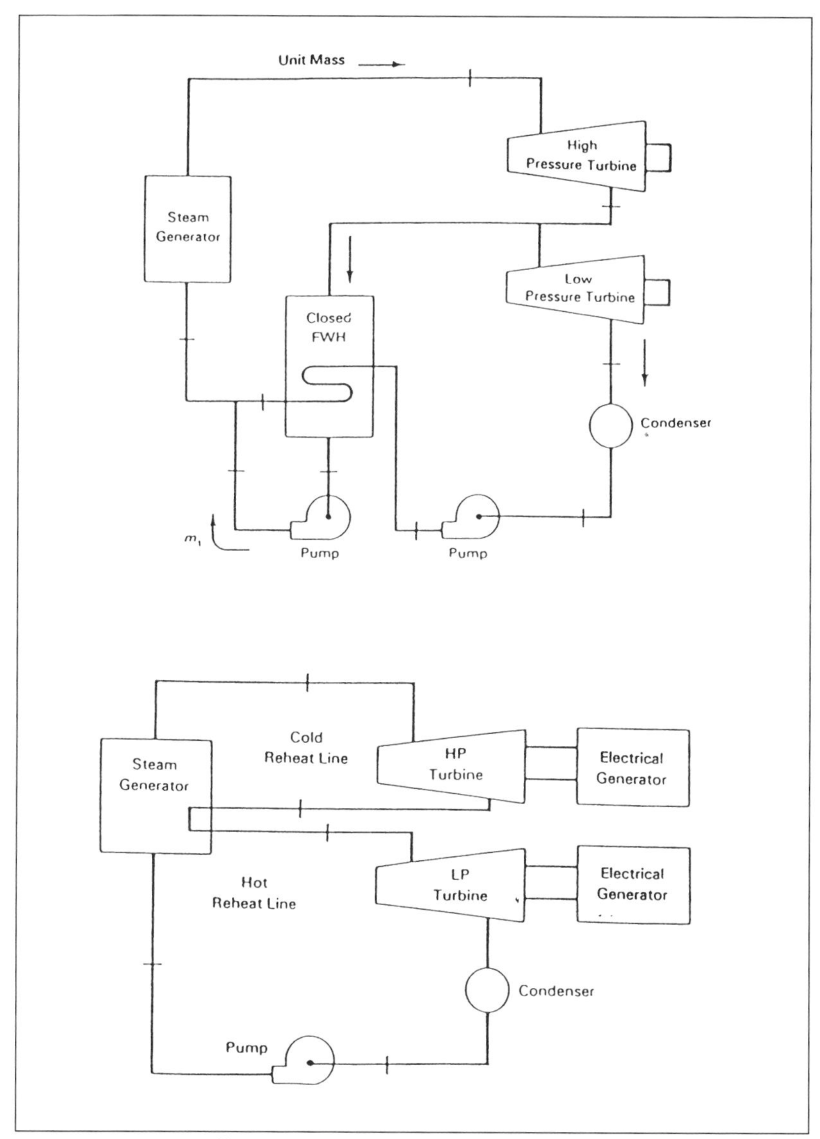

Figure 10-2: More-efficient Power Plant Cycles

Elements of a steam power plant

As noted in chapter 2, the generation of electricity is really the conversion of *kinetic* energy (geothermal, hydro, solar, etc.) or *potential* energy (coal, oil, gas, and uranium) into electrical energy. Fuel selection and

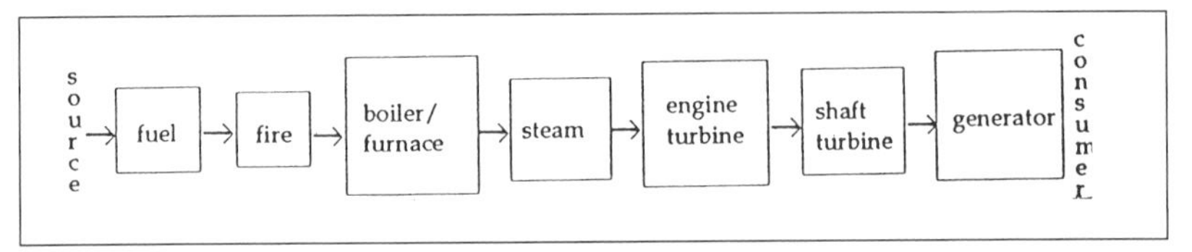

Figure 10-3: Schematic Diagram of Energy Conversion (Generation)

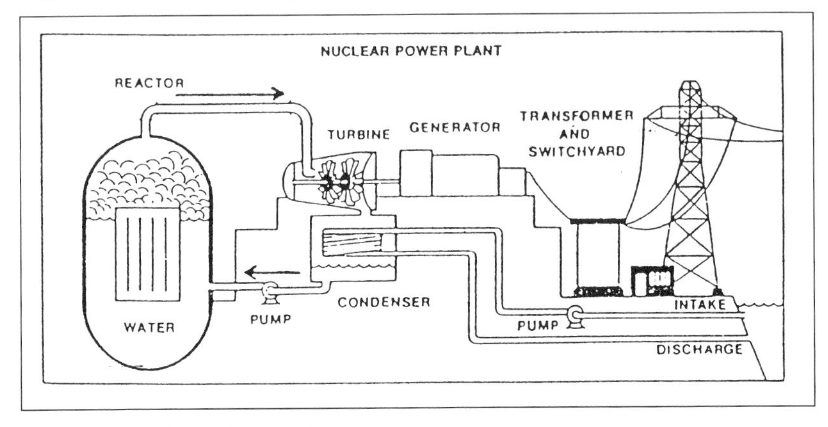

Figure 10-4: Electric Power Generation

handling are discussed in chapter 5. In chapters 6, 7, 8, and 9, we reviewed the elements that make energy conversion possible. Now it is time to put the pieces together. Our focus will be the classic steam-fired power plant fueled by coal. Nuclear and gas-fired plants are explored in succeeding chapters, though much of what is said here applies to them.

The rock-bottom basics can be seen in Figures 10-3 and 10-4. As potential energy (coal, petroleum products, nuclear fuel) is burned it is converted to heat energy (steam) and then into mechanical energy in an engine or turbine, and finally into electrical energy in the generator. Fuel burned in the furnace of a steam boiler produces steam that drives a turbine (or steam engine), which is connected by a shaft to the generator. Much of what is involved in this process is as old as the art of cooking—yet, technology continues to improve upon it. The entire process is classically divided into two operations—the furnace/boiler and the turbine/generator.

Maximizing combustion is the key to the furnace/boiler section. *Combustion*—a chemical process uniting the combustible content of the fuel with O_2 at a rapid rate—converts the fuel's potential energy into heat energy necessary to boil water for steam. *Complete combustion* means applying maximum amounts of heat to the boiler section for the most efficient production of steam. Note that we strive here for "complete" and not "perfect" combustion. *Perfect combustion* would mean that an exact amount of O_2 combines with all combustible constituents of the fuel with no fuel or O_2 wasted. Complete combustion means the total *oxidation* ("combining with O_2") of all combustible constituents with some amounts of O_2 left over (and lost as waste heat). Power plant operators are constantly seeking the Holy Grail of perfect combustion—and will never obtain it.

Combustion produces by-products. From coal, these include carbon monoxide, CO_2, H_2, water, sulfur, plus any number of other actors. What is important to our understanding here is the heat that is used to make steam in the boiler. "All heat is not equal," and we need the highest quality heat we can obtain. This is important because solid fuels such as coal do not directly ignite and burn; they must be heated to a temperature at which a portion gasifies. The gasified portion ignites and starts the burning process that delivers the heat we need. Because the greatest part of coal is carbon and the ignition temperature of carbon is 870oF, combustion will not occur until that ignition temperature is reached.

Once ignited, fuel will burn hotter than its ignition temperature. Should it fall below that set point, combustion will become erratic and eventually cease. That does not make for the high-quality heat our furnace needs to produce! To achieve consistent, maximum heat, we need to balance the rate of combustion—as hot as possible to achieve complete combustion but not so hot that the furnace and boiler are damaged. This balance is maintained by increasing or decreasing the amount of air (called *draft*) reaching the combustion chamber. Primary ignition to a coal bed is usually effected with a natural gas flame.

Boilers absorb all possible heat from their furnaces by means of water tubes circulating through the water drum in the lower half of the boiler. The water circulating in the tubes absorbs heat from the furnace, boiling the water in the water drum, and creating steam in the upper drum. The steam is collected and injected into the turbine, under pressure. Pressurized steam striking the turbine blades cause rotation of the shaft to which the blades are attached. This mechanical energy spins the generator, converting the mechanical energy into electrical energy—the point of the entire enterprise! (Figs. 10-5 through 10-15)

Figure 10-5: 2.3 MW Plant provides electricity and steam to the campus of the University of Massachusetts

Figure 10-6: GE 1.5MW Steam Turbine Generator

Figure 10--7: Powerboard controls power factor

Figure 10-8: "Igor—never touch this switch!" Safety systems are often very obvious

Figure 10-9: 5 Synchronizing with the grid requires constant monitoring

Figure 10-10: The exciter and…

Figure 10-11: Slip ring commutator on a turbine

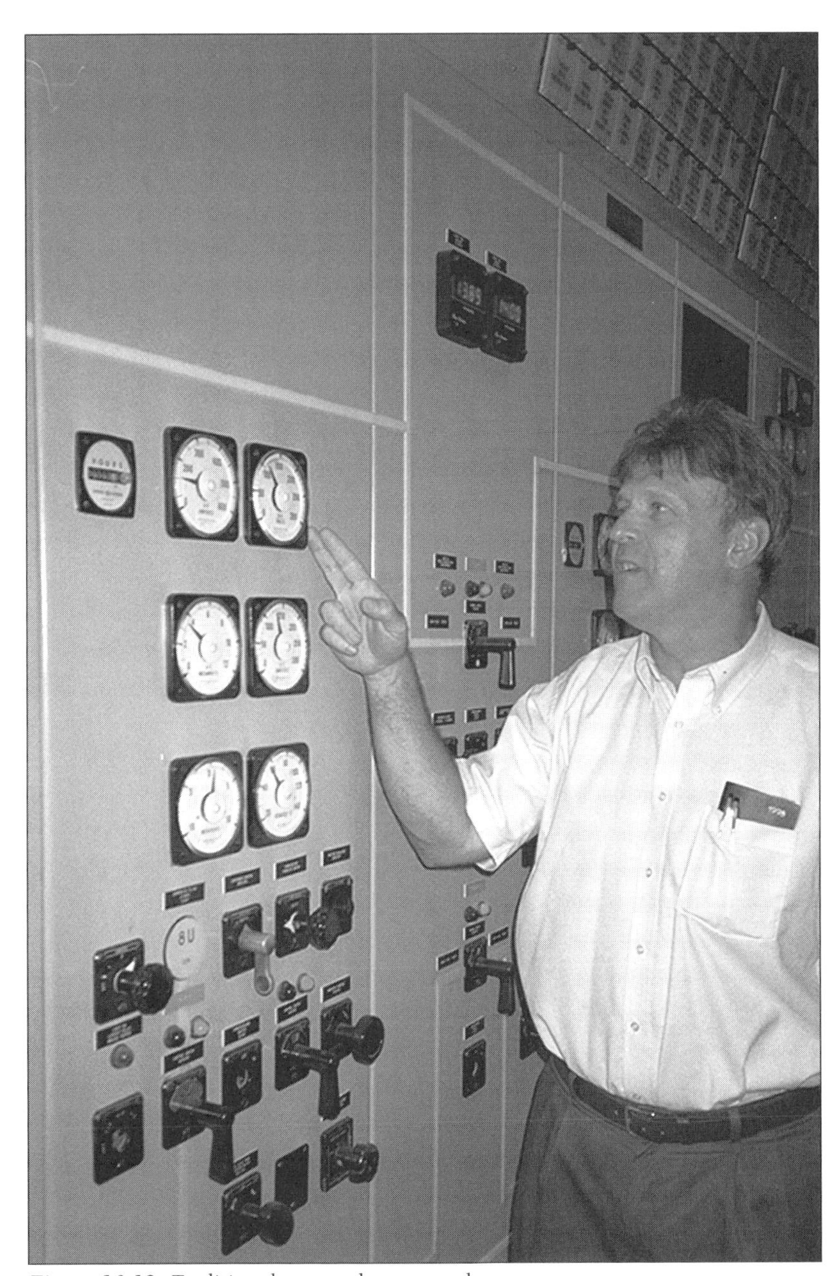

Figure 10-12: Traditional steam plant controls

Figure 10-13: 5 Computerized System Control and Data Acquisition (SCADA)

Chapter 11
Nuclear Steam Plants

"This is not really an art anymore. We gain 105 years of experience every year."

—Brian Naeck, Alternate Control Room Operator
on nuclear plants sharing information

This is a book of facts describing electric power generation, hopefully in an entertaining fashion. No other section in the book requires the kind of introduction that follows. The gentle reader is kindly asked to bear with us.

In 1939, at the urging of physicists Leo Szilard, Eugene Wigner, and Edward Teller, mathematician Albert Einstein wrote a letter advising President Franklin Delano Roosevelt of the potential of nuclear energy and proposing that the U.S. develop an atomic weapon ahead of the Nazis. Although Roosevelt ordered an effort be made to develop such a weapon, significant progress was not achieved until after the U.S. entered the war in 1942 with the establishment of the Manhattan Engineering Project.

The atomic genie was conjured under the direction of Enrico Fermi on December 2 1942, with the first controlled chain reaction. That genie was let out of the bottle with the first plutonium bomb test July 16 1945 and things have been confused ever since. That the genie even existed was the best-kept secret ever.

More to gain and more to lose

In the public's mind, the death toll and illnesses caused by the atomic bombings of Hiroshima and Nagasaki and atmospheric nuclear weapons testing run together with the nuclear power plant disaster at Chernobyl in the Ukraine and the perception of a disaster at the earlier TMI plant in Pennsylvania.

Nuclear energy was born in wartime secrecy and later ballyhooed as a source of electric energy "too cheap to meter." Because of the "mutually assured destruction" (MAD) mentality of the Cold War, the civilian nuclear power industry remained inextricably linked with the military and national security issues. In retrospect, decisions, policies and procedures were sometimes very flawed. Mistakes, personal wrongdoing and stupidity–the root cause of most conspiracies–could be covered up in the name of national security. These problems are still coming to light.

The mining and burning of coal takes more lives and causes more disease per kWh than comparable operations in the nuclear energy segment. However, as much as we recognize the advantages and disadvantages of coal as a source of energy, we deal with them more rationally than we do nuclear energy. Maybe this is because there is more to gain and more to lose with nuclear energy. Or maybe it is our fear of the unknown with a form of energy we cannot feel, hear, see, taste and do not really understand.

Those of us who do not work in the nuclear power segment of the electric power industry have our perceptions colored by the bombings of Hiroshima and Nagasaki and the movie *The China Syndrome*, which was released a week prior to the 1979 event at TMI. The 1983 film, *Silkwood* made the nuclear industry seem dangerous and overseen by very scary people.

Facts are still coming to light. They reveal that from the 1950s through the early 90s, civilians, soldiers, and workers were exposed to radiation from atomic weapons testing and nuclear fuel processing either intentionally to study the effects on humans or through neglect. At best, the abuses and neglect by the nuclear industry and its regulators in the wake of the war against fascism and then the cold war against communism reveal a government and industry supposedly based on enlightened self-interest had seriously compromised American values of honesty, human life, and justice. These problems were an insult and injury to the ingenuity, hard work, and honor of Americans from around the world—we were and still are a nation of immigrants—who in lab coats, coveralls, suits, and uniforms made both military and peaceful applications of nuclear technology possible.

The disaster at the Chernobyl nuclear power complex in the Ukraine permanently poisoned large areas of vital farmland and spread radioactive fallout over northern Europe and the western Soviet Union. The explosion of Chernobyl Unit 4 was only the latest of several serious accidents in the USSR which released radiation and contaminated the landscape. The immediate and short-term death toll from Chernobyl was variously estimated between 2,500 and 8,000. (The Canadian Nuclear Association states that the actual direct death toll was only 31. The much higher numbers are the total mortality of the 600,000 persons who worked on the decontamination and entombment of C4. These numbers are actually low for a comparable group of civilians.) Childhood leukemia has doubled in the affected region, but this amounts to hundreds of children at the present time, rather than thousands. Thyroid cancer, which is readily treated if diagnosed in time, is also up.

Long-term effects are seriously disputed. The cost of cleaning up contaminated areas and encasing the doomed Chernobyl reactor in concrete may have bankrupted the moribund Soviet economy and been the deciding factor in the downfall of communism in the Soviet Union. Soon, a new international effort to redo the reportedly crumbling "sarcophagus" surrounding the reactor containment may be required simply because neither Russia nor the Ukraine can afford to undertake it. While Reactor 2 was shut down after a fire on the conventional side of the unit in 1991, Reactor 1 continued to operate at Chernobyl.

Nuclear power has at times seemed too much trouble for the benefit received. However, through all the headlines, most nuclear power plants in the U.S. carry their share of the base load (about 15% of the total generation) without life-threatening incidents.

The aftermath of the atomic bombings of Hiroshima and Nagasaki did not result in the expected rate of birth defects and cancers. Fauna and flora on Bikini and other Aleutian atolls where American above-ground atomic testing took place returned to normal within a generation.

This is a book on electric power generation. We will stick to the facts here. We will not gloss over verified problems of the past. However, most serious problems in the nuclear industry have been in fuel processing, reprocessing, and weapons-related operations, rather than nuclear steam generating plants themselves. Indeed, Chernobyl was a fast-breeder carbon-pile, water-cooled reactor of the RBMK (*reactor bolshoj moshchnostij kanalnij* [high power channel reactor]) type producing weapons-grade plutonium for the Soviet military as well as electric power.

What are the facts?

The nuclear power industry began with the effort to build an atomic bomb in 1939. The U.S. Army took control in 1942, forming the Manhattan Project and proceeded rapidly. The first reactor to sustain a nuclear reaction was at the University of Chicago in 1942. This success led to building breeder reactors at Hanford, Washington to convert ^{238}U to ^{239}Pu. The Hiroshima bomb was made from the scarce ^{235}U isotope of uranium with a simple gun barrel design deemed so inherently reliable it was not even tested prior to use. The bomb dropped on Nagasaki used reactor-bred ^{239}Pu in an advanced implosion design tested at Alamogordo, New Mexico.

World War II atomic bombs and today's reactors work by splitting unstable atoms in a process called fission. Atomic bombs of the type used at the end of World War II must be more than 90% ^{235}U or ^{239}Pu. Reactor grade fuel is about 3% ^{235}U or other fissionable material such as plutonium or thorium. In atomic bombs, conventional explosive charges force subcritical masses of the fissionable material together at such a high velocity that a supercritical mass may be formed for an instant, with such an outpouring of neutrons it is followed immediately by a thermonuclear explosion.

In a reactor, the fuel rods are fixed in place and cadmium control rods, which absorb neutrons, are withdrawn to allow a sustained reaction to develop. Absorbing the neutrons effectively stops the chain reaction. In most power reactors, light (H_2O) or heavy (D_2O) water "moderates" (or slows) high-speed neutrons spontaneously emitted by ^{235}U nuclei making

Particle	Symbol	Charge e- = 1	Mass	Life in s	Comment
Alpha	a	++	7349	stable	He^2 nucleus. Blocked by paper or skin.
Beta	e-	-	1	stable	electron
Gamma ray		none	N/A	stable	Electromagnetic radiation. Shorter wavelength/higher energy than X-ray+
Neutron	n	0	1839	10^{10}	mass (relative charge to electron) in s
Positron	e+	+	1	stable	
Proton	p	+	1836	stable	

Table 11-1 Potentially Radioactive Particles

them more likely to split other ^{235}U nuclei (Table 11-1).

In a worst-possible-case scenario, cooling water could be lost while the control rods are for some reason stuck in the fully withdrawn position. Compare this to an automobile engine with the accelerator stuck to the floor and no coolant in the radiator. Eventually, the engine will throw a rod and the engine will seize. With the loss of the "water moderator" along with its cooling function, energy output would be reduced; however, the nuclear engine would not "stop." Intense heat concentrated in the fuel rods could eventually melt the fuel rods and cause them to collect on the specially designed bottom of the reactor vessel, further reducing the ability of emitted neutrons to split other nuclei. If nothing else went wrong, the reactor could safely exist in this mode indefinitely. Unlike bomb-grade uranium, fuel rods are only about 3-5% ^{235}U, so the fuel can be composed of chemically stable compounds such as uranium oxide that are unable to burn or react with other chemicals.

Two facts are worth keeping in mind when thinking about nuclear power:

1. The accident at the TMI nuclear power plant exceeded the "worst possible case" scenario for an event of its kind, but was still "within the design basis" for the reactor

2. Despite this, by any credible standard, radiation released during the event and the subsequent cleanup was insufficient to cause death or disease

Some physics

Nuclear power plants "burn" uranium, plutonium and/or thorium in the following reactions. Remember, what really goes on in a reactor is far more complex, but these simplified equations are adequate for our purposes:

Eq. (11.1) $${}^{1}_{0}n + {}^{235}_{92}U \rightarrow {}^{90}_{37}Rb + {}^{144}_{55}Cs + 2\,{}^{1}_{0}n$$

Eq. (11.2) $${}^{1}_{0}n + {}^{235}_{92}U \rightarrow {}^{87}_{35}Br + {}^{146}_{57}La + 3\,{}^{1}_{0}n$$

Eq. (11.3) $${}^{1}_{0}n + {}^{235}_{92}U \rightarrow {}^{72}_{30}Zn + {}^{160}_{62}Sm + 4\,{}^{1}_{0}n$$

Where,

$^{1}_{0}n$ = neutron

$^{87}_{35}Br$ = bromine isotope

$^{235}_{92}U$ = uranium-235 isotope

$^{146}_{57}La$ = lanthanum isotope

$^{90}_{37}Rb$ = rubidium isotope

$^{72}_{30}Zn$ = zinc isotope

$^{144}_{55}Cs$ = cesium isotope

$^{160}_{62}Sm$ = samarium isotope

Nuclear generating plants use enriched uranium, plutonium, and/or thorium in a controlled fission reaction to heat water until it turns into steam. Because the primary steam or pressurized water circuit is radioactive, in most nuclear generating plants a heat exchanger turns water in a secondary loop to steam, allowing the remainder of the plant to be operated in a conventional manner. The secondary loop has higher water pressure than the reactor side to prevent any leaks from contaminating the rest of the system.

Radioactive elements such as radium, uranium, plutonium—even the ^{14}C used to determine the age of ancient artifacts and carbon-bearing geological samples—have an unstable ratio of neutrons to protons in their nucleus. The number of protons in the nucleus determines the chemical properties of the element. In elemental form there is one electron for each proton. In most of the lighter elements, there is normally one neutron for each proton.

However hydrogen—the lightest element with one proton—breaks this rule before it even gets started. Hydrogen has three isotopes. The first is plain old hydrogen (H^1) with one proton, no neutron, and unless it is in its chemically stable gaseous H_2 molecular form, most of the time no electron. H_2 ions (H^+) in aqueous solution are what give acids their acidity.

The second isotope, deuterium (^2H or D), adds one neutron to the nucleus, seemingly following the one-neutron-per-proton rule. Deuterium occurs naturally but is extremely rare. D_2O is denser than H_2O and consequently has a slightly higher boiling point. This is exploited to extract it from water.

The tritium isotope of hydrogen (^3H or T) consists of a proton and two neutrons and is rarer still due to the fact that it is radioactive and has a half-life of 12.26 years. Tritium occurs naturally; however in 1954, the major source became atmospheric testing of thermonuclear weapons (^2H bombs). If all nations abide by treaties concerning nuclear weapons testing and the continuous warfare on this planet remains conventional, within a few dozens of years tritium levels in the environment will return to normal.

These isotopes are important in *fusion* reactions, in which ^2H nuclei are suddenly forced together under intense heat and pressure by a fission bomb to form helium atoms (releasing incredible amounts of additional energy in the process). Presently, fusion reactions are limited to thermonuclear weapons (hydrogen bombs). If a controlled fusion reaction can be harnessed to produce steam, fusion reactors promise inexpensive, plentiful energy with far less radioactive waste.

As noted above, in elemental form, each atom has one electron per proton. If "girls just want to have fun" as the song goes, atoms just want to have eight electrons in their outer shell. ^2H, with one proton, is happy to share an electron with almost any other element giving it a stable two electrons in its only shell. Helium has two protons and therefore two electrons. Helium does not need to react with anything to complete its shell-inert. Other light elements—lithium, boron, carbon, and others up to N_2, which have fewer than eight protons—must combine or react with other atoms to have their outer shell satisfied. Only in the inert gases such as helium, neon, krypton, xenon, etc. is this likely to be true. Hydrogen, oxygen, and nitrogen in their gaseous forms are most often found paired as ^2H, ^2O, ^2N, respectively. As an acid in solution (say, in your stomach) hydrogen is essentially a proton. Come 10 o'clock (or any other time), it has no idea where its electron or its chlorine ion is. If your acid indigestion feels like a nuclear reaction in your stomach, you are not too far from the truth. (Could reports of spontaneous human combustion actually be "cold fusion" events?)

Salt-forming elements such as sodium and calcium lend electrons to their counterparts such as fluorine and chlorine, forming what is called an *ionic bond.*

Elements combining with carbon and forming organic compounds tend to share electrons in *covalent* bonds to give each atom eight electrons in its outer shell. Covalent bonds tend not to conduct electricity well or at all. Metals tend to form lattice works for sharing electrons, which repeat endlessly up to the visible crystal level. This accounts for metals conducting electricity so well.

In all types of reactors, nuclear fuel becomes more radioactive as fission proceeds, not less. However, the unstable smaller elements produced emit particles other than neutrons, such as alpha and beta particles, which are not capable of sustaining a chain reaction. Only the ^{239}Pu created from ^{238}U in moderate amounts in standard reactors, and in large quantities in breeder reactors, is suitable for fuel.

Common elements

Common to all nuclear fission reactors built in the West (or of western design) are the following elements, from the outside in:

Containment building. Actually, this first item is not common to all nuclear power plants. If you can see a massive concrete, silo-shaped structure, then the containment itself is not contained.

Containment. A seriously steel-reinforced concrete building surrounding the reactor vessel. Access for refueling, and for passing control and instrumentation conduits and wiring is provided, but this structure is sealed to prevent radioactive materials from escaping during normal operation or in the event of a meltdown. The containment also houses the first heat exchanger separating radioactive water from the steam loop, which may directly connect to the steam turbine generator or to a second heat exchanger. Actuators for the control rods, equipment used during refueling operations, instrumentation, pumps, and other equipment required to control and monitor the reactor also reside within the containment. The containment below the reactor has the capacity to hold excess cooling water and safely hold melted fuel rods in a sub-critical geometry that would stop any nuclear chain reaction. Just as "scramming" the reactor by fully inserting the control rods is a real term in the nuclear industry, the containment would prevent the "China syndrome" in which melted fuel would bore its way "down to China."

Reactor vessel. Nuclear reactors operate at pressures up to 2,000 psi pressure, depending on type. The reactor vessel contains the fuel rod assemblies, cadmium control rods, and a "moderator"—usually water or D_2O—in a primary steam loop. The water either boils (as in a BWR) or is held at such high pressure (as in a PWR) that it doesn't boil. Either way, the primary loop of a heat exchanger isolates the radioactive water from direct exposure to the nuclear chain reaction occurring in the core, while conducting heat from the water.

As noted, water within the reactor vessel serves two functions. It "moderates" the fast neutrons into thermal neutrons that are more likely to split a ^{235}U nucleus, releasing heat and more neutrons. (Note that even in reactors not classified as "breeders," a neutron striking a ^{238}U nucleus may begin a process that leads to the creation of ^{239}Pu, which also is fissionable. In fact, by the end of a normal operating cycle in a BWR or PWR, up to 30% of the energy comes from plutonium fission.) The water also allows the reactor to do what it was built for in the first place—make steam to turn a turbine and run a generator. Unfortunately, "light" water (H_2O) does not just slow down electrons, it tends to absorb them. In Canadian deuterium uranium (CANDU) and other reactor designs, the D_2O moderator permits the use of unenriched uranium as a fuel.

The control rods used to stop the chain reaction are made of the metal cadmium. Cadmium was used for many years as plating on steel to prevent rust. Because it is a "heavy" metal and therefore somewhat toxic to animals, its use in consumer applications has been curtailed recently to reduce its presence in the environment. Cadmium is slightly less dense than silver, which has one less proton, and denser than zinc, which is above it in the periodic table of the elements. Both cadmium and zinc have two electrons in their outer shells and have similar chemical characteristics.

The fuel rod assemblies within the reactor vessel make it all go. Although the original Manhattan Project reactor built under the football stadium at the University of Chicago used metallic uranium (as well as uranium oxide for fuel) most reactors today use the oxide form. Oxides have the advantage of being stable chemically and highly resistant to heat. Tubes of zirconium alloys hold the uranium oxide pellets in the rod shapes used by reactors. Typically, the fuel rods are grouped into bundles and installed into the reactor. During refueling cycles, new fuel rod bundles are placed around the perimeter of the reactor core while older bundles are moved toward the center. The oldest fuel assemblies are removed for reprocessing or storage.

The only "problem" with the oxide form of fuel is that— like an old marriage that seems to break up without warning (uranium atoms are billions of years old, having formed in long dead stars, with ores at least millions of years old)—^{235}U nuclei can suddenly split into rubidium and cesium, bromine and lanthanum, or zinc and samarium nuclei. This produces between two and four neutrons, respectively, and a lot of energy—all with chemical properties significantly different from the original uranium. Therefore, the proportions of ^{2}O ions may be inappropriate to form new oxides. There may also be too much radioactivity and thermal energy in play to allow oxidation to occur.

Figure 11-1: Vermont Yankee Nuclear Power Plant. Photo courtesy of Vermont Yankee.

Boiling water reactor (BWR)

The nuclear power plant depicted in pictures throughout this book uses a General Electric BWR. While a BWR operates under considerable pressure (about 1,000 psi) unlike a PWR, some of the water is allowed to turn to steam along the fuel rods. Because turning a given amount of water into steam absorbs so much more heat than raising that water the last degree before it boils, steam conducts heat from the nuclear furnace more effectively than water. The water itself moderates the fission reaction, slowing down neutrons emitted by splitting ^{235}U nuclei, making them more likely to split yet another atom. If cooling water is lost, the high-speed neutrons are less likely to cause further fission. The water must be very pure to prevent mineral deposits from forming on the surfaces of the fuel rods when the water turns to steam.

Seen from the air, the Vermont Yankee Nuclear Power Plant can be identified most easily by the plume of water vapor rising from the cooling towers. The smaller capacity hydroelectric plant adjacent to it on the river stands out better in this lush rural setting (Fig. 11-1). The nuclear plant

Figure 11-2: Transformer Gives Access to 115kV Grid for Startup

connects three electric grids for generation and station service—a 345 kV transmission line to Northfield Mount Pumped Storage Station (shown in chapter 13); ISO New England down the Connecticut River Valley, and the 115 kV transmission system connected to the hydro plant (as well as its own internal 4100 V diesel generator backup system).

The fuel for nuclear power plants originated in long-forgotten star systems. After the laborious process of enriching typical uranium ores from 0.7% to approximately 3% of the ^{235}U isotope, it still takes energy to generate energy. Vermont Yankee connects its utility customers and other nuclear power plants in New England via 345kV transmission lines. However, it can start up from its 115 kV grid connection or its own diesel generators (Fig. 11-2).

While any generating station starts with a boost from the grid or its own auxiliary generators (just as the starter motor turns on an automobile) it is not the starter that fascinates us. Most of all, we want to get behind the steering wheel. In any power plant, it is all run from the control room (Figs. 11-3a and 11-3b).

Figure 11-3a: Alternate Control Room Operator Explains Logic Layout of Controls

Figure 11-3b: Computerized and Digital Equipment Supplement Original Analog Controls

Figure 11-4: To Control the Control Rods

Unlike a briskly efficient gas turbine plant, a comfortable old fossil-fired steam plant, and a laid-back pumped storage station, a nuclear power station is marked by armed guards at the gate, gracious, informative and unhurried tour guides, and very serious operators in the control room. The sense of competency and mission in the nuclear generating facility is impressive!

To operate the reactor at full power, all the control rods are drawn out fully. Figure 11-4 shows switches allowing the control rods to be individually positioned.

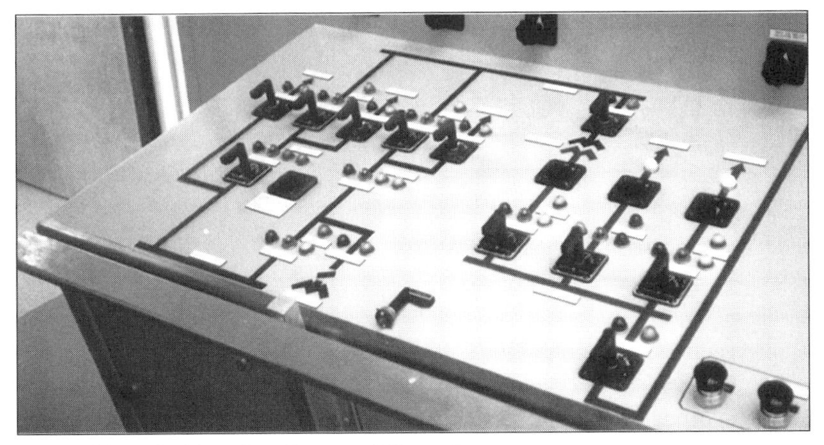

Figure 11-5: Grid Interface Status and Control

Figure 11-6: Output Status: 360kV and "Leading" Power Factor

For startup and operation, the plant must connect to one of the external transmission lines. Figure 11-5 depicts the section of the control console dedicated to grid interface.

The section of the control panel shown in Figure 11-6 displays output conditions indicating voltage of the plant at 360 kV, about 4% above the

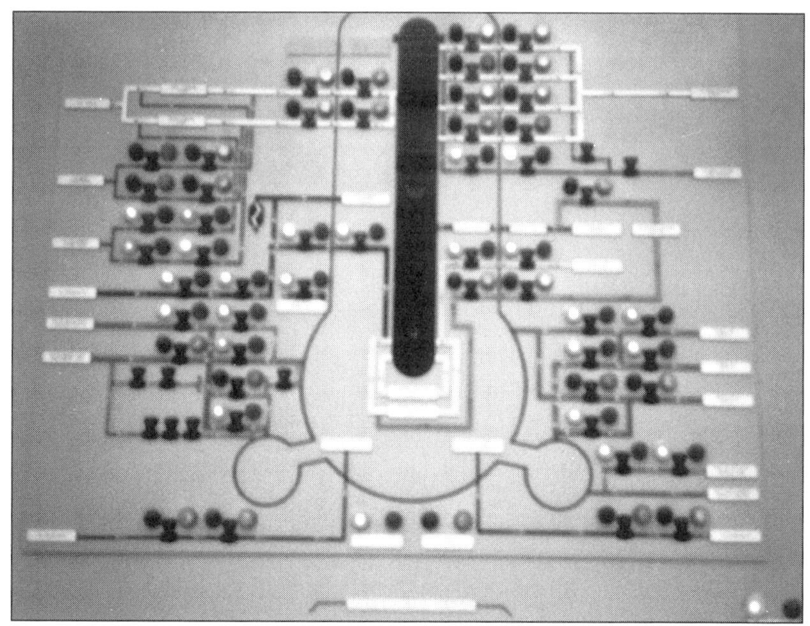

Figure 11-7a: Reactor Vessel Status

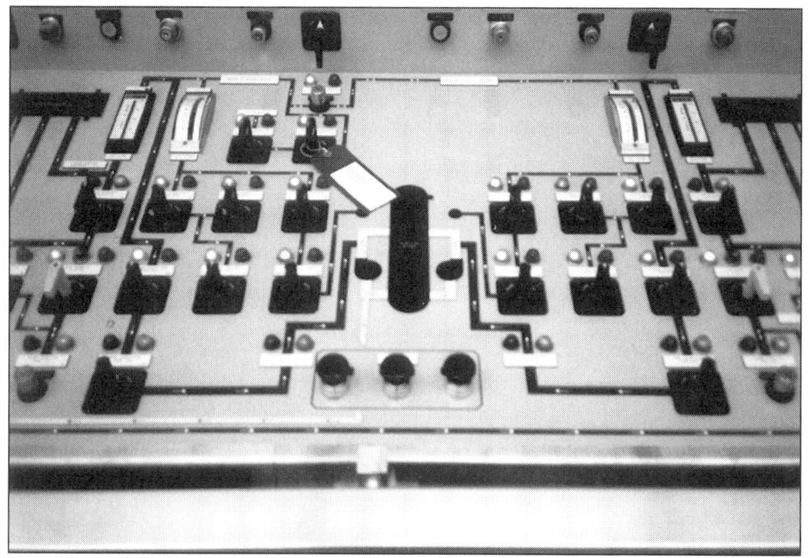

Figure 11-7b: Reactor Vessel Cooling

Figure 11-8: Computer provides powerful access and display of system status

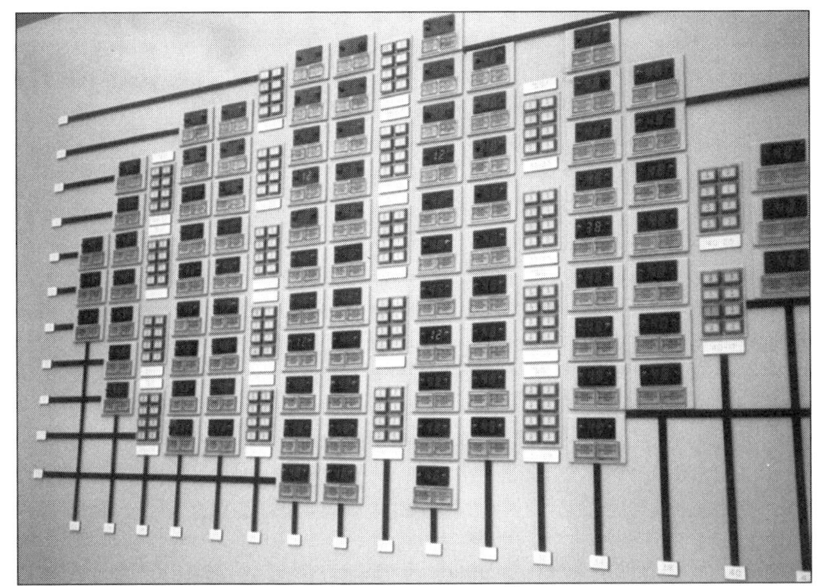

Figure 11-9: Reactor control rod position 48 = Fully Out

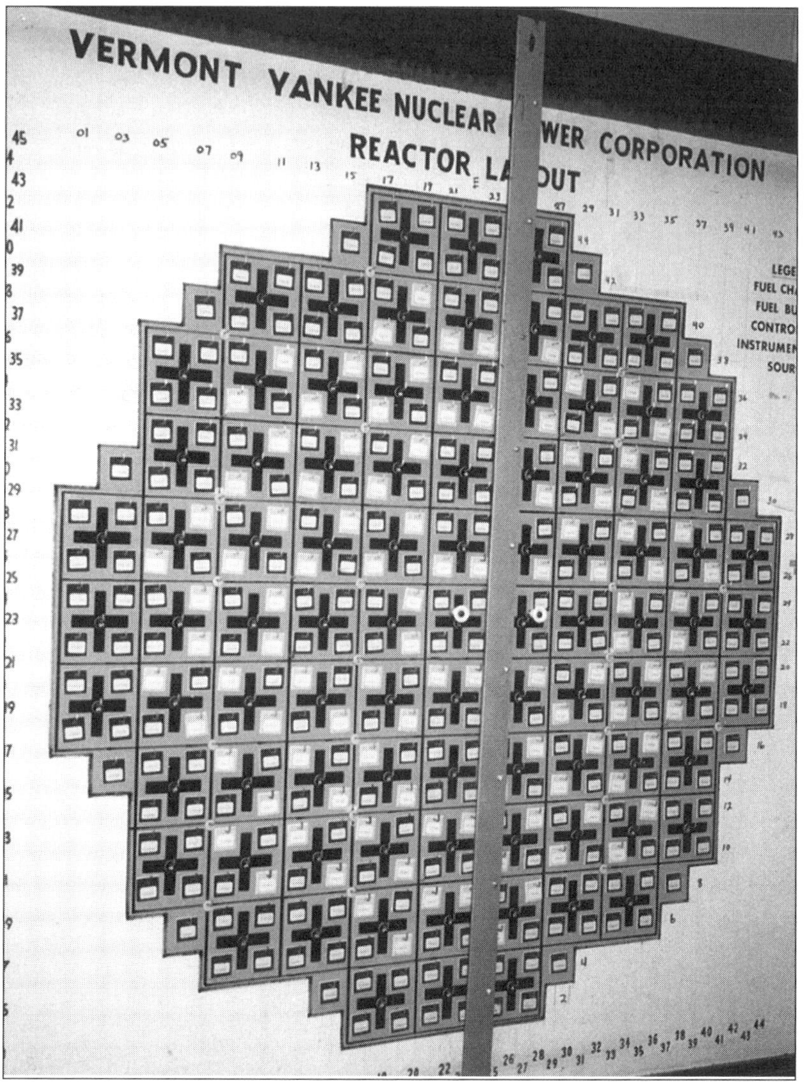

Figure 11-10: Reactor layout

nominal 345 kV line voltage and leading power factor. Ideally, generator and load would be purely resistive and the power factor gauge would be in the 12 o'clock position. In the real world, loads are normally reactive, usually inductive from motors and transformers. The generator must compensate, increasing the burden on the system.

Figure 11-11: Reactor instrumentation

Figure 11-12: Earthquake Reinforcing for Control Rod Actuators

Figure 11-13a: Primary hydraulic and backup control rod actuators

Figure 11-13b: Redundant control rod actuators

Figure 11-14: Pumps

Figure 11-15: Refueling Pool

Figure 11-16: Through the containment

Operators need to know exactly what is going on inside the reactor containment and vessel itself at all times, both during normal operation and "events" (Figs. 11-7a, 11-7b, 11-8, 11-9, and 11-10). With the pressures, radiation and heat involved—not to mention the thick concrete door blocking access to the containment—there is no going down to the reactor to check something you forgot. Figure 11-11 shows one of two virtually identical instrumentation racks that feed signals from sensors and transducers within the containment to displays and meters in the control room.

Not known for its earthquakes, like California, New England frequently has tremors too small to notice and an occasional one to rattle the China. However, some experts predict it is due for a quake that with its solid bedrock geology could make California look like small potatoes. Figure 11-12 shows steel framing to reinforce the reactor control rod actuators against earthquakes.

In a nuclear power plant, nearly everything has a backup—and usually a backup for the backup. Figures 11-13a and b show primary, backup, and redundant control rod actuators. To ensure critical supplies of cooling water are maintained, several sets of pumps circulate water through the non-radioactive heat exchanger loop (Fig. 11-14) during normal opera-

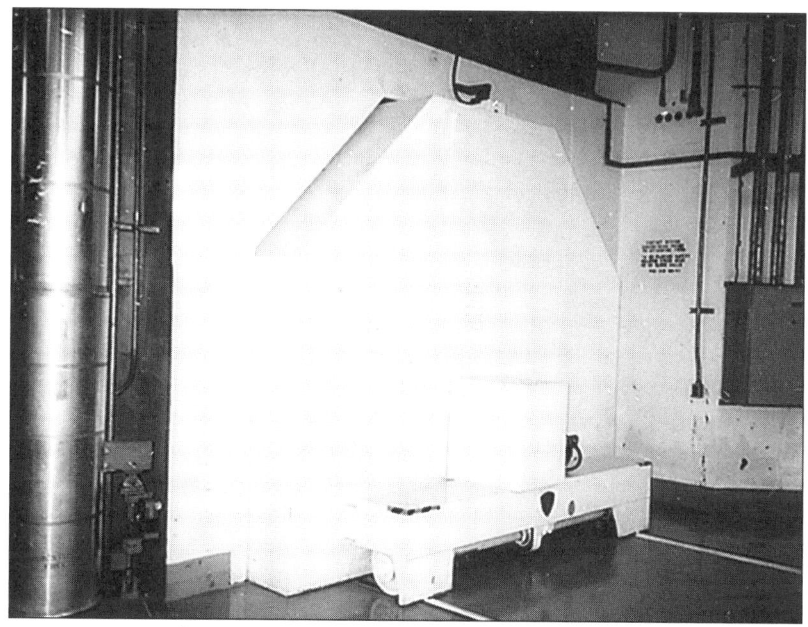

Figure 11-17: Refueling access

Figure 11-18a: Spent fuel pool

Figure 11-18b: Safe to swim in - but don't!

Figure 11-19: 0.000 mRem above reactor

Figure 11-20: Stack discharges low-level and treated waste

tion, and extra supplies of water are retained for use during refueling operations (Fig. 11-15).

Once a favorite entertainment, "The X-Files" lost much of its ability to suspend disbelief with an episode that showed a mutant creature seeking warmth by the reactor pool. As can be seen in Figures 11-16 and 17, access to the reactor is limited by instrumentation feed-throughs and a massive concrete door that is closed and sealed except during refueling operations.

Figure 11-21: "Turbine shine" gamma ray shield

Figure 11-22: Steam turbine

Figure 11-18a shows the spent fuel pool at Vermont Yankee. The pool in the containment building is much like an industrial grade swimming pool and is probably cleaner than your favorite pool. After many years of operation, only a small corner of the pool contains expended fuel rods and none has ever been removed from the site (Fig. 11-18b).

There is no measurable radiation released from a reactor during normal operation. In Figure 11-19, the plant's guide stands on the top of the reactor containment within the containment building and his electronic dosimeter reads 0.000 milliRems of radiation.

The multi-loop heat exchangers, thick concrete, and lead walls restrict radiation below natural background levels, which is vitally important—there is an elementary school across the street from Vermont Yankee (Fig. 11-20). Figure 11-21 shows the labyrinthine concrete walls that allow access to the steam turbine generator (Fig. 11-22) while absorbing minute amounts of gamma ray radiation (ionizing electromagnetic radiation with wavelengths shorter than X-rays) called "turbine shine."

Pure, dry hydrogen gas (H_2) molecules with both of their electrons make a low-viscosity, high-heat transporting, electrically insulating coolant for the 22 kV generator (Figs. 11-23a, 11-23b, and 11-24).

Figure 11-23a: Hydrogen Cooled Generator

Figure 11-23b: Generator

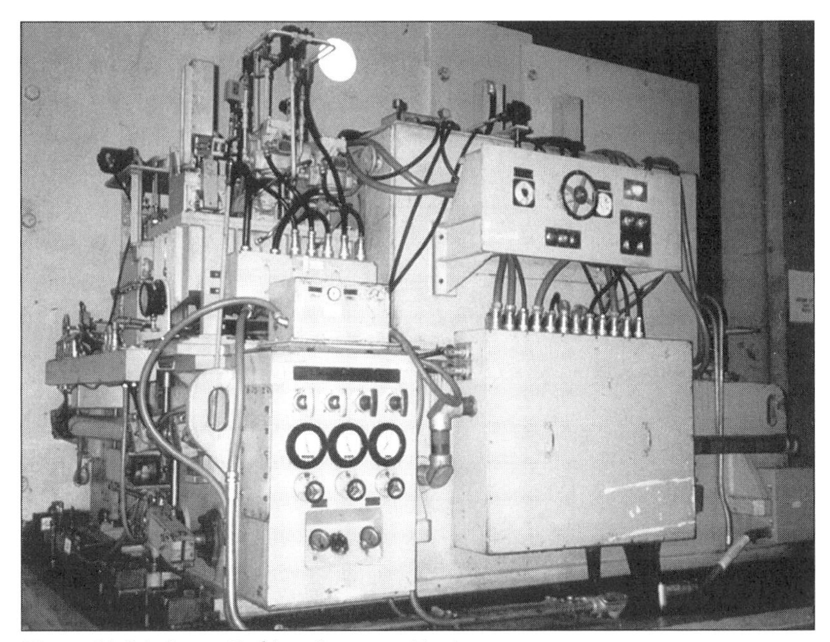

Figure 11-24: Steam Turbine Generator Exciter

Figure 11-25: Demarcation zone

As noted above, nearly all critical equipment in the containment building has backup and often redundant backup. Many times, it is physically separated on opposite sides of the containment. Figure 11-25 shows the "combustible free zone" that separates the two areas. Note the glossy floor

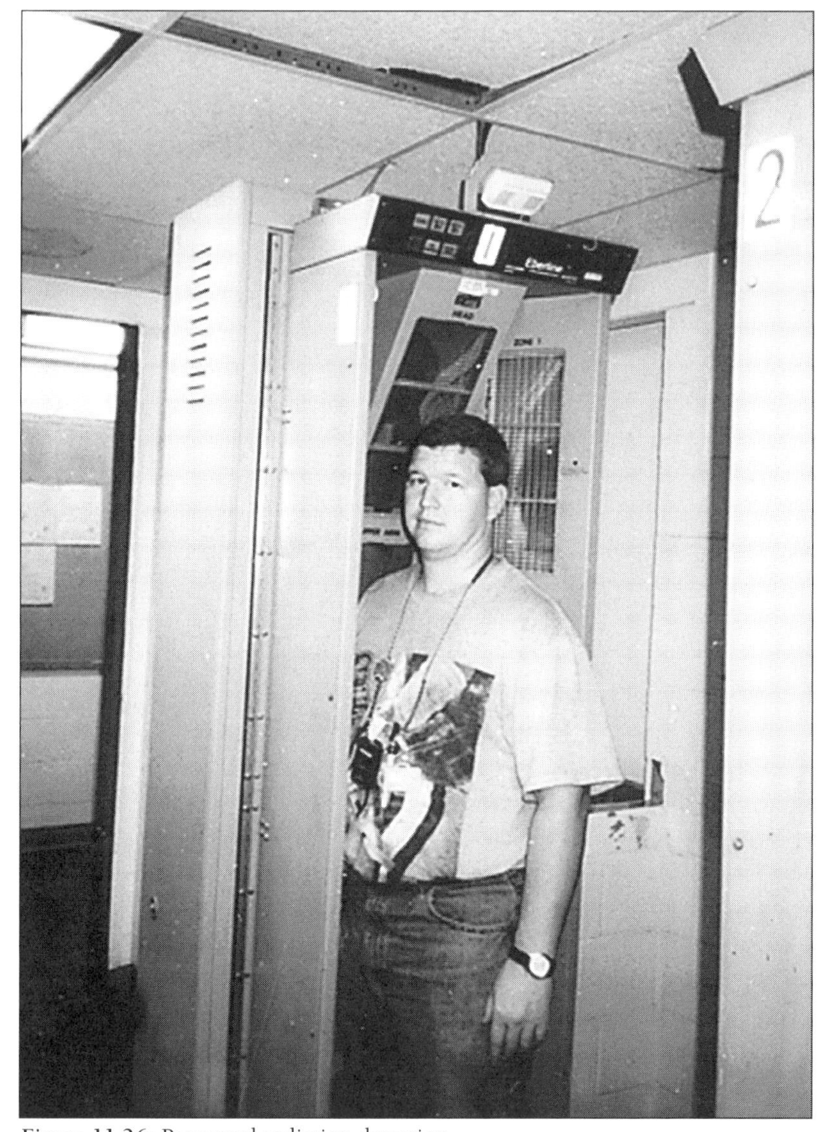

Figure 11-26: Personnel radiation detection

paint that prevents contaminants from sinking into concrete and making it permanently radioactive. Personal radiation detectors (Fig. 11-26) insure that at any time employees can be checked for radiation exposure.

Figure 11-27 shows what the excitement is all about: the main trans-

Figure 11-27: Main transformer steps up to 345 kV

Figure 11-28: Low-Profile Nuclear Plant Cooling Towers Surrounded by Trees

former steps up the 22 kV generator output to 345 kV for connection to the grid.

Waste heat is handled by two low-profile cooling towers. They are surrounded by trees so not to disrupt the picturesque Vermont countryside (Fig. 11-28).

Carbon pile reactor

The first reactor used in the Manhattan Project was a carbon pile reactor. Very pure carbon in the form of graphite moderates (or slows) neutrons. Unlike water, carbon does not have a tendency to absorb the neutrons, thus yielding the most effective chain reaction. Fast breeder reactors that create more fuel than they use turn common, stable ^{238}U into ^{239}Pu use carbon pile construction. Carbon pile reactors use cadmium control rods just as boiling water and PWRs do.

While very efficient on the nuclear side, graphite, like coal, in the presence of ^{2}O burns chemically exceedingly well, increasing the risk of a meltdown in the event of an accident that breeches the reactor vessel and allows in air. Reactors of this type have helium and/or ^{2}N atmospheres above cooling water. The reactors at the Chernobyl power plant in the Ukraine used carbon pile construction with a steel shell much weaker than the reactor vessels of western reactors.

CANDU reactors

As suggested by the acronymic, CANDU, Canada Deuterium Uranium reactors, developed in Canada, use D_2O as the moderator and natural 0.72% ^{235}U as the fuel. Because ordinary water absorbs about five times as many neutrons as deuterium, CANDU reactors spare the expensive and intensive process of enriching uranium ore to the 3% ^{235}U required by conventional reactors. D_2O is also laborious to separate from H_2O, but water is more plentiful and less toxic than uranium.

While reactors in the U.S. evolved from Admiral Hyman Rickover's nuclear Navy program and are all either BWRs or PWRs, the CANDU reactor claims benefits over these types. Relatively low tech in nuclear design but with more computer control, they offer less-developed nations access to nuclear energy without the need for enrichment plants that would tempt nuclear arms proliferation. CANDU reactors also claim more complete "burning" of nuclear fuel and up to double the life of fuel rods. Because the fuel rods may be changed while the reactor is on-line, no down time is specifically required for refueling.

While spent nuclear fuel is more radioactive than new, the half-lives of the various, newly created radioactive isotopes varies from seconds to years. Used fuel rods may be stored on-site in a spent fuel pool until radiation is low enough to allow them to be encased in a permanent container and transported to a long term storage facility.

Fast breeder reactor

Fast breeder reactors create more fuel than they use by exposing ordinary ^{238}U to neutron bombardment to become Pu^{239}. Breeder reactors are also used to produce weapons-grade plutonium for fission bombs. Fission bombs may be used by themselves or as the detonators for fusion "hydrogen" bombs.

A major advantage of breeding plutonium from ^{238}U is that it may be separated from the original uranium by chemical means. Isolating ^{235}U from ^{238}U is an arduous physical task based on the slightly higher density of ^{238}U.

Unfortunately, when plutonium spontaneously fissions, it gives off alpha particles. Outside the body, alpha particles are easily stopped by clothing, paper, or skin. If ingested or inhaled, plutonium-derived alpha particles are far more likely to cause genetic damage than smaller beta (essentially electron) particles or even gamma rays. Even though gamma rays have higher energy even than x-rays, their short wavelengths make them more likely to miss chromosomes in cells.

Liquid metal cooled reactor

Some fast breeder reactors have been cooled with liquid sodium. Because sodium is so reactive, it is never seen as a metal in nature. Artificially prepared sodium melts just below the boiling point of water at 97.8°C. It is not difficult to liquefy, and heat from the reactor will keep it liquid. While sodium transfers heat more effectively than water, it ignites and burns when exposed to ^{2}O. In the presence of water, it explodes. As long as everything operates normally, liquid sodium is the ideal coolant for a reactor.

Because breeder reactors have had so many functional and safety problems and water works so well both as coolant and moderator, it is unlikely liquid metal cooled reactors will see much application in the future.

Pressurized water reactor (PWR)

PWRs are the most widely used type in the West. Developed by Westinghouse for Admiral Hyman Rickover's nuclear Navy, PWRs quickly earned a reputation for being reliable and safe. (Whether this was true, or the PWRs were just as intimidated by the tyrannical Adm. Rickover as anyone else under his command, Westinghouse soon developed them for civilian electrical power generation use.) Other manufacturers soon followed, including Babcock & Wilcox (B&W), Combustion Engineering (now Asea Brown Boveri [ABB]), Framatome, Kraftwerk Union, Siemens, and Mitsubishi. The first nuclear submarine was Nautilus, SSN 571, launched in 1954, the same year the Soviet Union began operating its first nuclear power plant. Interestingly, the U.S. was actually the third country (after Great Britain) to produce electricity with nuclear power.

Very pure but ordinary water cools and moderates PWRs. Because H_2O absorbs about five times as many neutrons as D_2O, uranium ores must be enriched from about 0.7 to 3 to 5% ^{235}U to reach criticality. In all but the B&W vertical once-through design, three separate water loops conduct heat energy to run the steam turbine connected to the generator. In the B&W design, water does not stay in the reactor long enough to become radioactive and drives the steam turbine directly. In the other designs, the primary loop, which is highly radioactive from continuous recirculation within the reactor vessel, transfers its heat in an exchanger located within the reactor containment. Water in the secondary loop turns to steam to drive the turbine. A condenser on the low-pressure side of the turbine cools any residual steam turning it back to liquid water and pumps return the water to the heat exchanger within the containment. The third circuit provides cooling water for the condenser and may be a closed or open loop recirculating water to cooling towers or using water once-through from a lake, ocean, or river.

The pressurizer is unique to PWRs. It prevents boiling at the outlet temperature of 590° F by holding the water pressure at 2,250 psi. Unlike steam or any other gas, liquids, such as water, are incompressible. Because of the heat of vaporization of water, boiling conducts heat away from a superheated surface most efficiently. However, at pressures more than 1,000 psi higher than in a BWR, this benefit is minimal. Fluctuations in the nuclear chain reaction are eliminated because the moderator isn't constantly turning into an ineffective gas.

The ABB and GE advanced reactors have been certified for any future American installations. They are boiling water types. France continues to standardize on PWRs.

PWRs have viable competitors, but will continue to be the dominant type until the current generation of reactors is retired.

Advanced designs

Advanced reactor designs built around the world are not much used in the west. At this time, in the U.S., polls indicate the American public seems fairly receptive to nuclear power, but anti-nuclear activists have been successful in keeping edgy politicians from reopening the process for permitting, building, and operating new nuclear plants.

Advanced nuclear power plant designs include several important concepts:

- Simplified design: in an emergency, when all power is lost, a reservoir located uphill from the reactor provides cooling water
- Experience-based improvements: instead of adding redundant controls, instrumentation, and safety equipment as an afterthought, bottom-up designs improve reliability and reduce costs
- Standardization: all reactors will be of an approved and virtually identical design and execution so training can be standardized. There will be no surprises from reactor to reactor and plant to plant. Two advanced standardized reactors have been certified for domestic use—the GE Advanced BWR and the ABB Combustion Engineering System 80+.

Vive La Nuke

Ah, ze French...Americans love to return the disdain in which we perceive they hold us. Maybe it is author Barnett's Huguenot ancestry, or having worked for a French-owned company where visiting French managers clapped their hands and barked, "*Tout de suite!*" at female co-workers, which makes him inclined to indulge in a little French-bashing. It is not even politically incorrect to do so. We cite the arrogance they show Americans tourists.

"They started it, Mom!"

Arguably the greatest military general in history, Napoleon was a Corsican of Italian descent whose very success in conquering everything from Austria to Haiti to Holland to Russia to Sweden forged strange alliances among his enemies leading to inevitable defeat at Waterloo in 1815.

Despite the sinking of the British destroyer Sheffield in the Falklands Islands War by a French-made Excocet missile, the flourishing French military armament business typically sells to the runners-up—or worse, to both sides—in conflicts around the world. On the automotive side, how many people do you know who own or have owned a Citroen, Peugeot or Renault?

Without slighting our domestic industry, the French do produce the best wines in the world. French cuisine, traditional and *nouveau* is also a source of pride. However, other than food and wine, how does a nation perceived as mediocre to pretentious in industries similar and related to nuclear power generation rate a sidebar in this chapter? Is it their uniquely designed WMR (wine moderated reactor)?

"No gas, no oil, no coal, and no choice"

Could it be that despite the prevalence of rivers suitable for hydroelectric generation, as in the U.S., electric energy demand outstrips renewable resources by at least an order of magnitude? Could it be that French coal reserves are deep, difficult to mine, low grade, and have been easily annexed by Germany during past wars? Could it be that French natural gas supplies are high in sulfur and nearly exhausted in any case? Could it be that the French nation was highly motivated to achieve energy independence with innovative designs and a resolve to "do it right"?

We need to acknowledge that the French safely and successfully meet the energy needs of their country with limited natural resources using nuclear power.

France is the largest nation in western Europe, and third in size after Russia and the Ukraine in all of Europe. The land area is 210,026 square miles—larger than California and smaller than Texas. Hexagonally shaped, France shares borders with Andorra, Belgium, Germany, Italy, Luxembourg, Monaco, Spain, and Switzerland and has coasts on the Atlantic Ocean, English Channel, and Mediterranean Sea. The population is more than 58 million, with about 74% living in urban areas.

While no red-blooded American male would buy one of their cars, France produces the fourth highest number of automobiles and the third highest number of commercial vehicles in the world.

France is second only to the U.S. in the production of nuclear power. Eighty-one percent of their power comes from their standardized PWRs. Their nearly 60 plants (with others already under construction and coming on-line about one per year) constitute almost 60% as many plants as in the U.S.—and our number is static or dwindling.

Total French generating capacity is 105,520 MW with annual generation at 480,780 million kilowatt hour (MkWh). Fifteen percent comes from hydroelectric generation—comparable to the U.S.—but fossil fuels generate only 4%! Coal is certainly not king—nor emperor—in the French electric power industry. So how did the French achieve this nuclear miracle while the U.S. nuclear industry is mired in the radioactive ooze near the Paducah, Kentucky fuel processing plant?

Sparkling wine is valuable. Glow-in-the-dark wine would not be. Without any more French-bashing, it is fair to say France is more about good food, great wine, and Gallic pride than it is about electric power generation. Yet the industrial economy demanded a power supply that would not poison the Gallic Garden of not-quite-Eden, which is France.

As early as 1945, Electricité de France was charged with making France energy independent. In 1946, all electric generation, transmission, and distribution facilities were nationalized. By 1956, the French government ordered production versions of a prototype natural uranium-fueled, graphite-moderated, gas-cooled reactor designed by the Commissariat 'a l' Energie Atomique (CEA). Note that "natural" uranium is not something you buy in the organic section of your supermarket. "Natural" refers only to the 0.7% fraction of ^{235}U found in uranium ores. Because the graphite moderator slows, but does not absorb neutrons, "un-enriched" fuel converted to ceramic-like uranium oxides works just fine. One of the problems with graphite as a moderator is that it depends on the cadmium control rods operating to make the reaction sub-critical. Heavy (D_2O) or light boiling or PWRs inherently reduce power if coolant is lost. Graphite may also burn if the inert or non-oxidizing cooling gas escapes and ^2O enters the reactor. For these reasons, and others, in 1969 France began building

PWRs using light water (H_2O) as the moderator. The designs initially licensed by the American company Westinghouse and built by Framatome gave way to French designs in 1981. French companies COGEMA and Framatome are major players on the American and world scene, as well as in France.

The French nuclear power industry is horizontally and vertically integrated. Although conducted by several separate companies, these entities operate under government mandate. Electricité de France operates generation, transmission, and distribution equipment and outside-plant activities. COGEMA turns uranium ores into fuel, builds fuel claddings and assemblies, and reprocesses spent fuel. Ironically, plutonium is the most toxic component in spent uranium fuel and is readily removed by reprocessing. Mixed oxide fuels (MOX) are becoming the preferred nuclear energy sources for the new millennium. Fuel-grade concentrations of ^{239}Pu mixed with depleted uranium (mostly ^{238}U) makes more plutonium fuel in a conventional reactor.

France is the little nation that not only could, but had to and did. Nuclear power can be done correctly. Highly technologically capable countries like Germany and the U.S. have turned away from nuclear energy, but there is a successful model, should they choose to reevaluate the issue.

Chapter 12
Gas Turbine Plants and Cogeneration

In a gas turbine-powered electric generating plant, natural gas combustion turns the rotor of the generator. The prime mover in this case eliminates the "middle man" furnace and boiler.

Gas turbine plants are either simple-cycle or combined-cycle. *Simple-cycle* turbines pass hot gases through a compressor, combustor, and turbine. *Combined-cycle* turbines additionally capture heat from the exhaust gas and use it to power a second generator, heat water or generate steam for industrial applications.

Gas turbines are growing in popularity-especially in new construction-for one simple reason—*efficiency*. Fuel is the primary operating expense in any power plant and gas turbines allow plants to use the potential energy of the fuel with less waste. (This assumes generators have access to sufficient supplies of gas at affordable prices.) Gas turbines—and especially combined-cycle plants—are less expensive to build than coal, nuclear, or renewable electricity generation plants. Command and control are state of the art (Figs. 12-1, 12-2).

Clean fuel and increased efficiency also translate into low emissions. Electricity generation produces 35% of the nation's carbon emissions and

Figure 12-1: Computerized plant control

Figure 12-2: Computerized unit control

coal—accounting for slightly more than half of total U.S. generation—produces 87% of electricity-related carbon emissions. Gas-fired generation, which accounts for 15% of generation, produces only 9% of those elec-

Figure 12-3: "Smokeless" smokestacks. All three units are in full operation.

tricity-related emissions (Fig. 12-3).

Gas turbines were first placed into electricity generation service in the mid 1900s. The technology received a boost during World War II through Allied aircraft engine development programs, culminating in 1943 when Westinghouse tested the first modern American jet engine. After the war, utilities continued development and began installing gas turbines (officially known as *aeroderivative* gas turbines). The first power-generation gas turbine in commercial use was installed in 1949 at Oklahoma Gas & Electric Company's Belle Isle Station. Technology continued to advance despite the Fuel Use Act of 1978, which mandated discontinuation of natural gas as a fuel for electric generation because of perceived fuel shortages. When the legislation was set aside and the market returned in the late 1980s, manufacturers marketed even better turbines.

About 25,000 MW of gas turbine power is ordered across the globe every year, accounting for 35% of the world's fossil-fueled power plant capacity under construction at any time. Current simple-cycle units generate up to 150 MW with efficiencies around 35%, while combined-cycle units produce more than 200 MW with efficiencies approaching 60%—the highest efficiencies among all fossil fuel-fired power plants.

Figure 12-4: Gas turbine generator sets (like this one) use combustion gases to drive a turbine to provide electricity and recover heat from exhaust gases for waste-heat boilers.

A highly efficient turbine remains a challenge to engineers on a couple of fronts—but one with great rewards. Their compact dimensions (relative to other prime movers) produce a lot of energy for the their size and weight. Their ability to be operated on (comparatively) low-cost natural gas—as well as diesel fuel, naphtha, methane, crude oil, low-Btu gases, vaporized fuel oils, and even waste products—has made them the choice of the petrochemical and electric utility industries. (Fig. 12-4)

Their major limitations—and the focus of additional research—.are turbine inlet temperatures: they cannot always be maintained as hot as needed for most efficient heat rates. Research is concentrating on inlet design and metallurgy—of both the inlet materials and the turbine blades—to enable them to handle the higher temperatures. Regenerators heat compressed air as it enters the turbine to improve those heat rates. Because regenerators work from waste heat, fuel efficiency is extended. Combustors are used to increase the temperature of the high-pressure gases used on the fuel side.

Gas turbines, whether single- or combined-cycle, are either axial flow or radial flow in design. *Axial-flow designs* are used in 80% of turbine applications. "Axial" means the gas "flow" enters and leaves in an axial direction (along a single axis). It is best used for larger loads and over a large operating range, though the machinery is much longer than *radial-flow designs*. As the turbine extracts kinetic energy from expanding gases flowing from the combustion chamber, it converts that energy into "shaft horsepower" to drive the generator. Stationary nozzles flow the steam at high pressure onto turbine blades attached to the rotor to do so.

Cogeneration

Cogeneration puts into practice the old saying, "One man's trash is another man's treasure." In this case, waste heat from one industrial process is used as the source energy for another.

Specifically, cogeneration enables waste heat from the generation of electricity to power a second generator or to perform additional work—boil water, heat buildings, and power industrial processes. Conversely, "cogen" systems utilize waste heat from industrial or manufacturing processes to generate electricity. When electricity is generated first, the process is called a *topping cycle*. When waste heat from an industrial process is used to generate electricity, the system is called a *bottoming cycle*.

Because power plants waste as much as two-thirds of their thermal output, cogeneration can help maximize efficiency. It also can reduce amounts of certain airborne pollutants. When one fuel performs several functions, it requires less fuel to perform the same amount of work and reduces greenhouse gas emissions. Such systems have as advantage the fact they can be applied to both new construction and existing plants.

At the end of 1998, the Energy Information Administration of the U.S. Department of Energy (DOE) suggested that global generation of electricity from cogeneration was close to 200 GW—or 6.5% of the world's total generating capacity. U.S. capacity was also around 6%. It ranged from 60% in Sweden to 2.5% in France. Cogeneration facilities most often use gas turbines, diesel and gas engines, and fuel cells or microturbines.

The sensible use of cogeneration depends on several variables that need to be determined at the outset of any project. The overall question to be answered is this: is it most effective, from a cost standpoint, to serve a *thermal load* (the amount of work to be done with the waste heat) through cogeneration or some other heat source?

When to cogenerate

Prevailing electrical and gas energy costs, as well as public utility commission regulations have to be considered at the outset. Why install cogeneration systems if it is cheaper to buy gas and apply process heat, or if your state regulators make you jump through too many hoops? It is also useful to determine *where* a cogen project can best be located within a plant, and to do so even before process or facility heat needs are identified. Once those needs are subsequently identified, a reasonable balance can be struck between source and use.

A cogeneration package must then be considered in light of several essential factors:

1. Can the electricity generated by a cogen system be profitably sold to the grid or to a third party (or used internally)?
2. Is the installed cost of the system likely to be paid back in a reasonable amount of time?
3. Are the operating, maintenance, and personnel costs associated with the system within reason? This needs to include the cost of permits and inspections
4. If turbine exhaust is used to produce steam, what are the fuel savings over what would have been required to produce that same amount of steam in a boiler?

Capital costs are critical in considering any project. Because fuel and electricity rates fluctuate—sometimes greatly and not always in sync with one another—analysis of a cogeneration project's true value may be difficult to determine (risk factors). When we produce electricity but there is no ready use for the thermal portion of the process, *that* heat is wasted. If we need process or facility heat but not incremental electricity, *that* portion is wasted. Overall, the cogeneration system's *total utilization rate* has to be determined and evaluated for both heat and electric power. The best alternative may prove to be an intertie with the grid to constantly meet the heat load by taking or passing electric power as needed for balance. Systems operating less than full time (*i.e.*, seasonal heating use) have to be assessed differently than "24/7/365" projects. (Fig. 12-5)

Fuel prices and electric rates vary from region to region and so the wisdom of a cogeneration system may follow the realtor's maxim on property values: "location, location, location!" Generally speaking, the more stability found in fuel prices (especially natural gas), the better for the fuel/electric equation. What electric utilities are willing or able (or allowed) to pay for cogen-generated electricity also has to be evaluated.

As with all projects, solid engineering and sound construction and testing techniques are mandatory.

Figure 12-5: It's always raining inside cooling towers! Gas turbine cooling towers use the inherent efficiency of water's heat of vaporization to transfer heat just as effectively as boiling water in a steam turbine generator.

Chapter 13
Hydroelectric vs. Pumped Storage

Hydroelectric power is not just a "renewable energy resource." It is an energy resource that is continuously renewed every time it rains and when winter snows melt in the spring. A pumped storage facility may look like a hydroelectric plant, but it actually uses more electrical energy than it generates. Once converted from potential to kinetic energy, electricity is difficult to store efficiently. A reservoir may not be much more efficient than a battery, but does not require hundreds of thousands of tons of toxic heavy metals or harsh chemicals. Moreover, a battery does not get a free boost when it rains (Figs. 13-1, 13-2)!

Hydroelectric and pumped storage plants are prime movers that create no net heat load on the environment. The falling water would give up its energy as heat from friction as it passed over riverbanks and bottoms, rocks, in turbulent flow, and from evaporation. No matter how precise and well lubricated bearings may be, where metal spins against metal—whenever electric current flows through wires and when magnetic fields build—"local" heat is produced.

Hydroelectric power *does* have an environmental impact, however. Damming rivers for hydropower affect river flows and impacts events such

Figure 13-1: Dam, hydroelectric plant, and mill canal, Turners Falls, MA

as spring flooding. Originally seen as a benefit of taming rivers, limiting spring runoff can cause lakes or rivers above dams to fill with silt, which would naturally fertilize agricultural bottomlands below the dam. The dams also prevent migrating fish from reaching their spawning grounds. Recently plans to simulate natural spring flooding on some rivers have been discussed. "Fish ladders," permitting salmon to swim upstream so they can spawn in their natural habitat, have been built at some hydro plants. Hydropower plants may significantly impact surrounding areas, as well, if their reservoirs cover up towns and farmland.

Hydroelectric plants are best sited along fast-moving rivers or streams, in mountainous regions, and in areas of heavy rainfall. Of the nation's 80,000 dams, 2,400 are used for hydropower. Many other dams could be retrofitted to produce electricity and even more hydroelectric power could be generated without building dams in the first place.

History and development

From antiquity, people used the mechanical energy in flowing water to operate machinery-originally to grind grain and corn. The word "mill"

Figure 13-2: Hydroelectric plant II

originally meant the machinery for grinding corn or working metal. Quickly the term grew to include the building or complex housing the mill or mills. Wheels using waterpower to drive belts and pulleys or rotating shafts adapted easily to early generators.

Soon water turbines designed specifically for hydroelectric generation evolved and the characteristic vertical-shaft generator was born. The first hydroelectric power plant was built in 1882 in Appleton, Wisconsin to provide 12.5kW to light two paper mills and a home. By the time electricity emerged from laboratory experiments and "parlor" tricks in the Nineteenth Century to become a practical source of energy and illumination, water and steam power were the two most efficient methods of generating electricity. While not the biggest, the first, or the most difficult hydroelectric project, the incredible technological advancements of the Nineteenth Century culminated in large-scale power generation at Niagara Falls.

Hydropower has played a major role in making electricity a part of everyday life and helped to power industrial development. It currently produces almost one-fourth of the world's electricity. Today's hydropower

Figure 13-3: 60 kVA hydroelectric generator

plants range in size from very small to capacities up to 10,000 MW. Worldwide, hydropower plants have a combined capacity nearing 700,000 MW and annually produce over 2.3 trillion kWh of electricity. With a capacity exceeding 92,000 MW—electricity enough to supply the needs of 28 million households—the U.S. leads the world in hydropower production. Hydropower supplied 304,403 MkW/h in 1998, the majority of all renewable energy used in the U.S. The largest hydropower plant is the 6,480 MW Grand Coulee power station on the Columbia River in Washington State.

Basic hydropower

Hydropower converts kinetic energy in flowing water into the mechanical energy of the water wheel, which turns a generator to produce electrical energy (Fig. 13-3). How much electricity is produced depends upon how much water is flowing and how far it falls to strike the waterwheel—called the "head distance." The greater the flow and the "head," the more electricity that is produced. The entire package—from water inflow to generator—is considered a prime mover in a hydroelectric plant.

Hydroelectric plants impound (store) water behind a dam. Intake systems—canals, flumes, and pipelines, either natural or man-made—lead impounded water to the penstocks (Fig. 13-4). Penstocks are a series of pipes designed to balance surge and water pressure reduction as the water enters the valves (or gates) to reach the water wheel (or turbine). Water wheels are attached to generator rotors.

Water wheels are either reactive or impulse design. Reaction wheels use water pressure and the reactive force of curved blades much like propellers. The impulse wheel is designed to resemble a steam turbine: high-velocity jets of water strike buckets on the rim of the water wheel and turn it. Reaction wheels are best suited to combinations of low head/large water quantities; impulse wheels do best with high heads/small water quantities.

In place of the exhaust stream in a fossil-fuel plant, hydroelectric plants employ tailraces to carry spent water to the watercourse below the dam.

Two kinds of hydropower plants are generally recognized.

Conventional. These hydropower plants (which make up the majority of such facilities) use a one-way flow of water to generate electricity. Conventional plants are further divided into *run-of-river* and *storage* plants. Run-of-river plants use limited amounts of stored water to provide turbine flows. These plants can experience significant fluctuations in power output in cases of drought or seasonal changes to river flows. Storage plants maintain water supplies to offset such occurrences and so provide a more constant supply of electricity. (Figs. 13-5, 13-6)

Pumped storage. After making electricity, water in these facilities flows from the turbines into a reservoir below the dam and is stored. During periods of low electricity demand, the water is pumped into an upper reservoir and reused during periods of high demand. Again, these facilities actually use more electrical energy than they generate!

Hydropower plants can exist along side conventional fuels plants. (Fig. 13-7). Most hydropower plants are built by federal or local government agencies as part of multipurpose projects. Dams and reservoirs provide flood control, water supply, irrigation, transportation, recreation, and refuges for fish and birds, as well as electric power. Overall, they provide inexpensive electricity with no pollution. Unlike energy produced with fossil fuels, water is not consumed and can be reused for other purposes.

Figure 13-4: Draft tube normally full of water

Figure 13-5: Hydroelectric dam on Connecticut River at Turners Falls, MA., provides flood control, electric energy, and water for mill canal. Fish ladder reduces environmental impact.

Figure 13-6: Extreme turbulence from hydroelectric outlet into mill canal makes for good fishing at Turners Falls dam.

Figure 13-7: The Vernon hydroelectric plant is downstream from the Vermont Yankee nuclear power plant.

Chapter 14
The "Alternative" Fuels: Geothermal, Wind, Solar, Tidal, and Biomass Generation

> Alternative: (adj.) "Existing or functioning outside the established cultural, social, or economic system;" (noun) "A proposition or situation offering a choice between two or more things or courses."
>
> —Webster's Collegiate Dictionary, 10th edition

Electric power generation's *establishment* fuels can be said to include coal, natural gas, and nuclear resources (although as recently as a generation ago nuclear was considered to be the new kid on the block). *Alternative* energy sources, for our purposes, embrace geothermal, wind, solar, tidal, and biomass. Hydropower—a source of energy as old as the rivers and the ancient peoples who first harnessed them—can rightly be considered an establishment fuel. However, as a non-polluter, it is also "claimed" by the proponents of renewable-resource power generation.

This raises an issue that needs immediate clarification: "alternative" in this reference has nothing to do with the social and political upheavals in the 1960s! A distinct philosophy of power generation (and consumption) evolved from the era—that is true. It can generally be summarized as environmentally conservative and decentralized, clamoring for increased use of renewable energy sources and decreased energy consumption across all residential, commercial, and industrial sectors. For our purposes, however, "alternative" simply means technologies for the generation of electricity other than the classic fossil fuel-fired generating plants. Our planet, growing as it is in population and sophistication, needs all the power it can generate and all prudent, proven technologies need to be explored and used.

Geothermal Energy

Geothermal energy refers to heat locked beneath the earth's crust and brought to the surface as steam or hot water when water flows through heated, permeable rock. It can be used directly for space heating or converted to electricity.

Geothermal energy is as powerful as the earth is big—but not very easy to access. The crust of the planet averages depths of 30 kms with a thick layer (mantle) beneath that is made up of a molten rock called *basalt*. It bubbles to the surface (or erupts from volcanoes) as lava (or magma) but otherwise cushions the crust. The source heat is the earth's core—apparently generating heat via some sort of radioactivity. We can tap this heat energy source only where it naturally occurs.

Geothermal resources are located in the western U.S. and provide more than 2,700 MW of electric power capacity from naturally occurring steam and hot water. Potential electricity production from other geothermal resources has to await the day we can figure out how to drill for it. Geothermal resources come in five forms—hydrothermal fluids, hot dry rock, geopressured brines, magma, and ambient ground heat. Only hydrothermal fluids have been developed commercially for power generation. It is still too difficult to make commercial use of volcanoes except as tourist attractions.

Converting hydrothermal fluids to electricity depends on whether the fluid is steam or water, as well as its temperature. Conventional turbines use hydrothermal fluids that are wholly (or primarily) steam, routing it directly to the turbine to drive an electric generator. This eliminates boilers and conventional fuels to heat water. Hydrothermal fluids hotter than 400°F are sprayed into a tank at pressures lower than the fluid, causing some of the fluid to rapidly vaporize (flash) to steam. The steam drives a turbine and a generator in a conventional fashion. Liquid that remains in the tank can be flashed again for more power generation. When water is less than 400°F, the hot geothermal fluid vaporizes a secondary—or working—fluid, which then drives a turbine and generator.

While steam resources are easy to use, they are rare. The only commercially developed steam field in the U.S. is the Geysers in northern California. It began producing electricity in 1960—the first source of geothermal power in the country. It remains the largest single source of geothermal power in the world. Hot water resources are more common than steam and provide the major source of geothermal power in both the U.S. and the world. Hot water plants operate in California, Hawaii, Nevada, and Utah.

These plants release little or no CO_2 and are very reliable when compared to conventional power plants. New steam plants at the Geysers operate 99% of the time. In some areas, geothermal systems compete with conventional energy sources on cost. An average geothermal system generates electricity at 5 cents/kWh to 8 cents/kWh.

Wind

Would it surprise you to learn that wind energy is derived from solar energy?

As the sun heats the earth, it converts thermal energy into kinetic energy—the five distinct wind systems sweeping across the planet. This occurs because the sun heats the earth's surface and atmosphere unevenly and heat transfers from warmer to cooler areas of the planet. These thermal differences kick up the winds. The earth's rotation helps to keep air currents fluctuating, as well. It is a constant seeking of balance between thermal and mechanical forces that will never be attained.

Windmills have been used to tap into this energy from the time of the ancient Chinese, Greeks, and Egyptians, to power machines used to grind grain and draw water. Industrialization triggered the development of large windmills (actually wind turbines) for the generation of electricity. The

Figure 14-1: Feathered experimental wind turbine generator

major difference is that faster windmills are necessary to generate power than to draw water. Commercial development of windmills fluctuates with the price of fossil fuels. Interest in them fell when fuel prices fell after World War II—when oil prices vaulted in the 1970s, so did interest in wind turbines. Most wind-powered systems in use today were established between 1970 and 1984 when federal tax credits essentially jump-started an industry. The credits were discontinued in 1985 and development was scaled back. However, the legacy of those years is improved efficiency and affordability of wind turbines, so the cost per kWh is close to that of conventional fossil fuel generation. Wind-generated electricity that costs more than 13 cents/kWh in 1984 has fallen to around 5 cents/kWh (Fig. 14-1).

The U.S. has more than 1,883 MW of installed wind capacity to meet the annual residential needs of a million people. Almost all of it is produced by wind farms in Altamont Pass, Tehachapi, and Palm Springs, California.

Modern wind turbines convert the wind's kinetic energy first into mechanical energy and then into electric energy using airfoils, a rotor, and a generator. Air flowing over a blade (an airfoil) rotates the blade around a rotor, which connects to the generator. The airfoil is not unlike an airplane wing that provides lift, except that where the moving airplane uses engine power to "create" wind (and lift), the stationary airfoil uses the wind to create power.

Wind speed is the critical factor. Because the amount of power available in the wind is proportional to the cube of the wind speed, relatively small changes in wind speed result in large changes in power. In addition, the wind's power varies from place to place and from season to season, so in choosing a productive site for a wind system meteorology plays a role.

To compensate for erratic winds, technology has enabled longer airfoil blades to reach out and grab more power. Size is limited only by factors such as material, joint strength, weight, and cost. Blades approach 100 feet in length and wind turbines now in use in the U.S. can generate up to 1,300 kW of power.

The power produced by a wind turbine—its rated power—is what a turbine should produce when run at its rated wind speed. Winds that are consistently faster than the rated speed wear out turbine components and cause maintenance headaches. Braking systems are installed to lessen the aerodynamic nature of the blades, slowing them down, and mechanical systems turn blades somewhat away from the wind or slow the drivetrain.

Wind generation systems can be sized and installed to match a given load, from a single small turbine to a wind farm. Although they must be spaced to avoid diluting the wind's power, turbines allow the ground on which they stand to be used for agriculture or cattle grazing. Emissions are not a problem—no cooling towers or smoke stacks. However, wind turbines are noisy and are not a welcome sight in residential communities.

Figure 14-2: Neglected "solar house" at the University of Massachusetts, Amherst.

Environmentalists have also complained that birds die when they fly into the blades of wind turbines.

Solar

Each day more solar energy strikes the planet than the total amount of energy the planet's inhabitants could consume in three decades. If the solar energy falling on an area 700 miles square were converted into electrical energy, it would equal the peak generating capacity of every power plant on earth. Is this a great source of power, or what?

Unfortunately, a combination of high capital costs, geographic realities, and technology shortfalls mean that solar energy is not going to reach its potential anytime soon. This far in its development, solar power-photovoltaic and thermal systems-are best utilized for specific assignments (hot water heating for laundromats, powering hand-held calculators, motion-detected security lighting, etc.) and not for replacement power for the grid (Fig. 14-2). Solar-based generation is not even as dependable as distributed generation resources because until viable electrical storage systems are developed, solar does not cut it in the dark.

Though all regions of the planet receive some amount of sunlight, the amount varies so much (depending on location, time of day, season of the year, and weather conditions) that solar heating or power generation is not practical in very many places. The southwestern U.S. is one very good place. This desert region receives twice the usable sunlight of other areas of the country. But because so few people live there, its benefits are limited. The cost of production of solar equipment dropped by 75% between 1975 and 1980 but has not moved very much since then, despite advancements in more efficient photovoltaic cells and better mass production techniques.

Solar energy systems use either solar cells or some form of solar collector to generate electricity or heat homes and buildings. The primary solar energy technologies for power generation are photovoltaics and thermal systems. Photovoltaic devices (solar cells or modules) use semiconductor material to directly convert sunlight into electricity. Power is produced when sunlight strikes the semiconductor material and creates an electric current. Light energy is really an energy stream called photons. When they strike the solar cell, they generate an electron flow in the semiconductor material (which has extra electrons). The light energy and solar cell materials set up opposing positive and negative charges and current flows. This current is drawn off into electrical circuits. Each cell produces about a half a V dc (.1 to 4 amps) and solar cells must be joined with hundreds of other cells in solar arrays to provide a usable amount of current for grid-based systems.

Solar thermal systems generate electricity with heat. Solar collectors use mirrors and lenses to concentrate and focus sunlight onto a receiver mounted at the system's focal point, which absorbs and converts sunlight into heat. The heat is sent to a steam generator or engine to be converted into electricity. Solar thermal electric systems are of three types:

- parabolic trough collectors that use mirrored troughs to focus energy on a fluid-carrying receiver tube located at the trough's focal point
- parabolic mirrors that concentrate and focus incoming solar energy onto a receiver mounted above the dish at the focal point
- central receivers that use thousands of individual tracking mirrors (heliostats) to reflect solar energy onto a receiver located atop a tall tower

Tidal

We know from experience with hydropower generation that moving water represents a great deal of potential energy. We know, too, that water displaced by ocean tides represents enormous amounts of energy. Tidal energy systems, while still experimental and obviously relegated to coastal locations, may represent tomorrow's power source.

Most designs call for the erection of dikes that fill at high tide, holding water when tides recede. (Power could be generated by inflowing waters as well.) The water behind the dike represents the potential energy and as it is released through a hydropower system (operating at "very low head"), it can be converted into electrical power. The obvious drawback is that power can be generated only twice a day (when tides recede) and only areas with extreme differences in tides will suffice.

If currents flowing across the world's oceans can be analogous to the earth's wind systems, then perhaps this potential energy could be tapped, as well. The problem would be what to do with electrical energy generated in the middle of an ocean!

Biomass

Biomass is organic matter that is burned in an incinerator and converted into a combustible gas to produce electrical energy. Fuel sources include agricultural, forestry, and food-processing byproducts as well as the methane gas emitted from landfills. The electrical energy is derived from the potential energy locked in plants and organic matter. In addition to generating electricity, biomass can heat homes and provide process heat for industrial facilities. The U.S. currently derives 3% of its energy needs from biomass resources.

Unlike coal, oil, uranium, and natural gas that have to be mined and drilled for, biomass is literally all around us. Principal resources include wood and wood wastes, crops and their waste byproducts, municipal solid waste, animal wastes, food processing wastes, and aquatic plants (including algae). Most commonly used resources are wood and wood wastes, municipal solid waste, agricultural waste, and landfill gases. Dedicated energy crops—grasses, grains, and trees grown specifically for energy production—are expected to make a significant contribution in the next few years.

So is the federal government. In August 1999, President Clinton ordered the EPA and the departments of energy and agriculture to work with industry and farming representatives to triple the use of bio-based products by 2010. The aim is to increase farm income, boost rural communities, and reduce greenhouse-gas emissions, in addition to reducing dependence on fossil fuels. Research grants and tax credits are proposed.

At this writing, wood-related industries and homeowners consume the most biomass energy. The lumber, pulp, and paper industries supplement their energy needs by burning wastes in large furnaces and boilers. Wood stoves and fireplaces for cooking meals and warming residences are popular in some 20 million homes.

On a smaller scale, biomass is used to generate electricity and manufacture liquid and gaseous fuels and a variety of chemicals currently manufactured from petroleum. The major drawback is that its energy is less concentrated than the energy released from fossil fuels and so technologies must be developed to make it competitive with them. Vegetation-type biomass feedstocks are used in power plants the same way fossil fuels are. Landfill gas plants that collect methane gas (the primary component of natural gas) can run generators.

Generating electricity from biomass can be accomplished in several ways. By direct combustion, biomass is burned to produce steam that turns a turbine and a generator. Potential ash build-up and maintenance expense means that only certain biomass materials can be used in direct combustion. Gasification processes convert biomass into a combustible gas (biogas) that is used to drive a gas turbine. Pyrolysis is a heat process by which biomass is chemically converted into pyrolysis oil, which is burned to generate electricity.

Biomass energy generates far fewer air emissions than fossil fuels, reduces the amount of waste sent to landfills, and decreases our reliance on foreign oil. To increase use of biomass energy, dedicated energy crops must be developed, system efficiencies must improve, an infrastructure to efficiently transport biofuels must be developed, and the cost of biomass energy must become cost competitive with fossil fuels.

Biomass is also converted into transportation fuels such as ethanol, methanol, biodiesel, and additives for reformulated gasoline. So-called biofuels can be used in pure form or blended with gasoline.

Ethanol, the best-known biofuel, is made by fermenting grains in a process similar to brewing beer. Ethanol made from corn and blended with gasoline helps to improve vehicle performance and reduce air pollution. Methanol is biomass converted into a synthesized gas (syngas) similar to natural gas. It is used as a solvent, antifreeze, and is synthesized into other chemicals and blended into reformulated gasoline and racing fuel. Biodiesel fuel is made from oils and fats found in micro-algae and other plants, and can be substituted for, or blended with diesel fuel. Biomass is also used to produce reformulated gasoline components such as methyl tertiary butyl ether (MTBE) or ethyl tertiary butyl ether (ETBE).

Chapter 15
The Future: Building on What's Been Done

There are many ways to produce electricity. Electrons flowing between different materials provide a current in a common chemical battery, i.e., a reliable, portable, power source that also runs down quickly. Unfortunately, a battery is not suitable for use in the large generating plants that have helped power the industrialized world since 1945.

To provide the steady power demanded by modern societies, large power plants have been built to produce massive amounts of electricity with a machine known as a generator (more properly called an alternator). Essentially, an alternator consists of a rotor (an electromagnet that rotates) and a stator (one that remains stationary for easy connection to the station transformer). The principle of electromagnetic induction makes generation possible. Passing a wire through a magnetic field generates a current in the wire. Wind that wire around an iron core and the voltage at the opposite ends of the wire multiplies according to the number of turns. All alternators, including the one under the hood of your car, produce ac. In your car's alternator, diodes-electronic one-way valves-immediately convert the output to dc compatible with the battery and electrical system.

The key to understanding alternators is that the prime mover turning the generator provides the mechanical kinetic energy, while the current used by the rotor coil only establishes the magnetic field allowing the conversion to electrical kinetic energy.

All generators—including Edison's original dynamos—are devices in which the magnetic field is established in the stator winding and the generated electricity comes from the rotating winding. In small generators, permanent magnets similar to the ones holding the art gallery and schedules on your refrigerator may be used instead of an electromagnet. Conducting the generated electricity from the rotor via a pair of brushes and slip rings in an ac generator (or in a pair of brushes and many pairs of commutator contacts in one of Edison's dc dynamos, or dc automobile generators through the early 1960s) wasted a lot of precious power as heat, and required frequent maintenance.

It is important to recognize that electricity is not mined or harvested. It must be generated. This will not substantially change! Except as a static electric charge and in superconducting experiments, electricity cannot be stored and must be manufactured at the time of demand. Electricity is a form of energy, but not an energy source. Different generating plants harness different energy sources to make electric power. The two most common types are thermal plants and kinetic plants.

Thermal Generating Plants

Thermal plants use heat energy to make electricity. Water is heated in a boiler until it becomes high-temperature steam. This steam is then channeled through a turbine, whose fan blades are attached to a shaft. As the steam moves over the blades, it causes the shaft to spin. This spinning shaft is connected to the rotor of a generator, and the generator produces electricity.

Fossil-fueled plants

Fossil fuels are the remains of plants and animals that lived long ago. Decomposing in swampy peat bogs, eventually silted over and exposed to high temperatures and pressures for millions of years underground, these remains have been transformed into forms of carbon—coal, oil, and natural gas. Unlike electricity, fossil fuels can be stored in large quantities. After 100 years of research and development, fossil-fueled plants are generally reliable, and problems that do occur are usually confined to a local area. Many electric utilities have operated fossil fuel plants for decades, and these plants (now fully paid for) are very profitable to run. This not only increases profits to the utility, but keeps down the direct cost to users.

However, the chemistry of fossil fuel combustion causes various environmental problems. Depending on the type and grade, burning these fuels produces SO_2 and NO_x air pollution requiring expensive scrubbers. Wastewater from the used steam can carry pollutants into watersheds. Even with very good pollution controls, there is still waste material produced. CO_2 gas and ash are the current concerns although there are new studies that suggest the CO_2 produced is absorbed by trees, making this problem a potential wash. In addition, fossil fuels are not (immediately) renewable. They took millions of years to make and once they're gone, they're gone for an equivalent period of time. Extracting and transporting them for use can also create environmental problems. Strip mining of coal and oil spills at sea have adversely impacted ecosystems, as well.

Cogeneration

As oil has become too expensive for use as a fuel for power plants, and as prices plants pay for coal and natural gas have declined in terms of real dollars, utilities throughout the U.S. are using this latter group more often. Coal and natural gas are being used most efficiently in cogeneration plants.

Cogeneration is a way to produce electricity that begins not with the generating plants but with the customers. It takes advantage of the way many large electricity users operate. Many factories use steam in their production process. Utilities often make and sell steam for these customers, as well as running their own generators. Rather than simply condensing and exhausting waste steam after it has passed through the turbine, topping cycle cogenerators pipe this usable commodity to nearby customers. Bottom-cycle cogenerators operate in reverse. They use waste steam from industrial processing to drive turbines and make electricity.

Note that in gas turbine cogeneration plants, the jet-engine-on-a-stick always comes first. In these plants, topping cycle means the superheated exhaust gases from the gas turbine first boil water to run a steam turbine generator. The residual steam then is used for district heating or industrial processes. In a bottoming cycle gas turbine cogeneration plant, the exhaust gases from the gas turbine make steam for heating or industry and then any residual steam turns a steam turbine. By reusing steam, thermal efficiency at cogeneration plants can exceed 50%. In a sense, cogeneration plants provide "free" electricity or steam, depending on how you look at it.

Recently developed cogeneration plants have improved reliability and controlled both thermal and atmospheric pollution. Since these new technologies are designed into plants from the start, they are less expensive to install. The economy and capability of cogeneration technology allows many plants to return to burning coal without exceeding air-quality standards. Circulating fluidized-bed boilers, selective catalytic (and noncatalytic) reduction, and zero-discharge water-treatment systems are examples of technologies being used to control various environmental problems.

Combined-cycle and biomass plants

Some natural gas plants produce electricity using turbines very much like those on jet aircraft in place of steam boilers. Instead of burning jet fuel and producing thrust, these units burn natural gas and turn generators. Gas-turbine generators have been popular for many years because they can be started quickly in response to temporary demand surges for electricity. A new twist is the combined-cycle plant that uses gas turbines in this fashion, but then channels the hot exhaust gas to a boiler that makes steam to turn another rotor in the conventional way. This substantially improves the overall efficiency of the generating plant.

In addition to these innovations, some thermal plants are designed to burn biomass. The term applies to waste wood or some other renewable plant material. For example, during part of the year Okeelanta Cogeneration Plant in Florida burns bagasse-waste from surrounding sugar-cane processing operations-and waste wood during the growing season.

Nuclear plants

Nuclear power stations make electricity the same way as fossil fuel plants do. The difference is that they generate steam by using the heat of atomic fission rather than by burning coal, oil, or gas. The steam then turns a generator as in other thermal plants. Nuclear plants don't use large amounts of fuel and do not refuel often, unlike a coal plant that must have trainloads of fuel shipped in daily. The fact that greenhouse gases and airborne particulates are minimal during normal operation makes nuclear power attractive to many who are concerned about air quality. Wastewater is hotter than that from a fossil plant, and large, distinctive cooling towers are designed to address this problem.

Despite their advantages over fossil plants, and their popularity around the world, the drive to field nuclear power in the U.S. faltered in the face of public concerns regarding safety, environmental, and economic issues. As more safety mechanisms were specified, construction costs and system complexities grew. Also, plants have shown some unexpected quirks, such as boiler tubes wearing out prematurely.

Experience has led to better and simpler designs with inherent safeguards. Two new advanced boiling water reactor (ABWR) plant designs have been certified in the U.S. Although American construction is indefinitely on hold, these reactors are being accepted around the world. The CANDU and French advanced pressurized water reactor (APWR) designs are also being installed in energy-starved, less-developed nations and technologically advanced ones such as Japan.

Opponents argue that using uranium and plutonium as fuel create problems and risks that outweigh benefits the technology might have. So far, one problem that has not been solved is that of disposing spent fuel cores and contaminated accessories that may remain dangerous for thousands of years. Permanent burial in geologically stable locations is the plan being pursued at this time, though this is still very controversial. High-profile accidents at TMI, in Pennsylvania, in 1979 and Chernobyl, in the old Soviet Union, in 1986 were public relations disasters for the nuclear industry. Continuing economic problems have made nuclear plants much less attractive investments. Even though it produced 22% of America's electricity in 1996, nuclear power's future in this country is uncertain and hotly debated.

Kinetic Generating Plants

Hydroelectric plants and windmills convert energy into electricity using kinetic energy—the energy of motion—instead of heat energy. Moving wind or water spins a turbine that spins the rotor of a generator. Since no fuel is burned, no air pollution is produced. Wind and water are renewable resources and, while there have been many recent technical innovations, have a long history of being harnessed as energy sources. The overall availability of renewable energy resources limits their value and they are not without environmental impact.

Hydroelectric plants

Two basic types of hydroelectric plants are in service. A "run-of-river" plant takes energy from a fast moving current to spin the turbine. The flow of water in most rivers can vary widely depending on the amount of rainfall. Hence, there are few suitable sites for run-of-river plants. Most hydroelectric plants use a reservoir to compensate for periods of drought and to boost water pressure in the turbines. These man-made lakes cover large areas, often creating picturesque sport and recreational facilities. The massive dams required are also handy for controlling floods. In the past, few questioned the common assumption that the benefits outweighed the costs.

A special type of hydropower is called pumped storage. Even some non-hydro plants can pump water into a reservoir and, when demand rises, channel it through a hydro-turbine to generate electricity. Since peak load generating units (such as small diesel, natural gas, and oil-fired plants) are generally more expensive to run than base load units (which run most of the time), pumped storage is one way to boost system efficiency and meet temporary demand surges.

Wind power

Wind farms do not need reservoirs and create no air pollution. Small windmills can even provide power to individual homes. Air carries much less energy than water, however,and so much more of it is needed to spin rotors. One needs a few very large windmills, many small windmills, or many large ones to operate a commercial wind farm. In any case, construction costs can be high.

Like run-of-river hydro-plants, there are a limited number of suitable locations where the wind blows predictably. Even in such sites, turbines often have to be designed with special gearing so that the rotor will turn at a constant speed in spite of variable wind speeds. Some cite less technical problems—installations that can turn a scenic ridge or pass into an ugly steel forest, or that can take a toll on birds, for instance.

Alternative Generation

Other types of power plants do not use traditional equipment to produce electricity. Geothermal plants replace boilers with the earth itself. Photovoltaic (PV) systems and fuel cells dispense with turbogenerators entirely. These alternate energy technologies have been under development for several decades and advocates believe the technical and political situation will now bring them into the marketplace.

Geothermal plants

Pressure, radioactive decay, and underlying molten rock make the deep places in the earth's crust hot indeed. A vivid example of the heat available underground is seen when geysers erupt, sending steam and hot water high in the air. Natural sources of steam and hot water have attracted the attention of power engineers since early in this century. By tapping this naturally created thermal energy, geothermal plants provide electricity with low levels of pollution.

There are several different varieties of plants, some of which produce products used for heating as well as electricity. Finding suitable sites can be difficult, although as technical innovations occur, more sites are made practical. Tapping geothermal sources can also have the effect of "turning off" natural geysers, and this possibility must be taken into account during the planning stage.

Solar power

Solar cells (or PV cells) do not use a generator—they *are* the generator! Usually arranged in panels, these devices take advantage of the ability of light to cause a current to flow in some substances. Photons—packets of light energy—raise electrons out of their normal orbits along a semiconductor junction. When the electrons return to their normal energy level, the PV cell conducts electrical energy.

Photocells wired in a series to form solar batteries produce no pollution. Most scientists predict that the fuel supply—the sun—will last at least 4 billion years. Solar panels have been relatively expensive to make, and of course they do not work at night or in foul weather. Like other semiconductors, photocells consist overwhelmingly of silicon, the most abundant element in the earth's crust. Small amounts of potentially toxic metals such as arsenic and gallium must be diffused in gaseous form at high temperatures into the silicon, and industrial handling of these materials has recently been called into question. Only a small fraction of the sunlight striking a solar cell is converted into electricity, and boosting efficiency has been slow work. Yet, the idea of harnessing all of that free sunlight remains a powerful driver for solar power.

Probably PV and thermoelectric technology need to be combined into a hybrid cell to significantly increase efficiency. Just as with conventional peaking generating plants, overriding need, rather than short-term cost effectiveness, will drive development of this and comparable technologies.

Fuel cells

Valued for their usefulness on spacecraft, fuel cells chemically combine substances to generate electricity. While this might sound very similar to a battery, fuel cells are powered by a continuous flow of fuels. In the U.S. Space Shuttle program, for example, fuel cells combine H_2 and O_2 to produce water and electricity.

Fuel cells have generally been expensive to make and not well suited to large installations. However, they represent a "modular" technology in that capacity can be added in small increments (5-20 MW) as needed, allowing utilities to reduce both capital expenditures and construction lead times. Research seems to show promise, too: one test installation in Yonkers, NY, can produce 200 kW by using gas created in the operations of a wastewater treatment plant. Also, fuel cell plants are being used for central power in Japan. If electric cars become economically viable, it will be because the energy density in the fuel compares well to the internal combustion engine. As in the space program, the best fuel for a fuel cell is

H_2, but alcohols work almost as well. Alcohols and H_2 can be at least partially obtained from renewable sources.

Decentralized generation

The ultimate usefulness of fuel cells or PVs—and even gasoline- or diesel-powered generators—may not lie with large central generating plants. In the era before great continent-spanning networks of wires, a small generating station on the premises made economic sense for many business and industrial power users. As motors and equipment were improved and designed to take advantage of the new energy supply, more customers electrified their businesses and homes. In the early Twentieth Century, small generating companies consolidated and independent plants slowly disappeared. It simply became more economical to purchase power from a centrally located utility rather than generate it on-site. Large regional power pools grew, as companies interconnected their transmission systems and shared reserve capacity. Economy of scale became the guiding principle.

This may change in the Third Millennium. As the technology of electrical generation improves, and environmental concerns rise, the very concept of large centralized generating stations is coming into question. In most cases, for example, it is not economical to heat homes and businesses from a central location. Individual furnaces provide heat for separate buildings, with fuel provided by an associated transportation and distribution system. Gasoline- or diesel-powered generators provide decentralized power to buildings in emergencies, though they are not economical for full-time power. Continued technical improvements in fuel cells or PVs may change these economics. This possibility is especially attractive considering the cost of and objections to building large power lines.

Cold Fusion Confusion

"Arthur C. Clarke, distinguished author of science and fiction, says ideas often have three stages of reaction: first, 'It's crazy and don't waste my time;' second, 'It's possible but it's not worth doing;' and finally, 'I've always said it was a good idea!'"

-former President Ronald Reagan to the National Space Club, March 29, 1985

In a recent on-line interview to inaugurate his foundation's website, Arthur C. Clarke—science fiction writer, scientist, inventor of the geostationary communication satellites, and possible inspiration for the Internet—reiterated his belief in cold fusion.

Clarke is a member of the cadre of engineers and scientists who helped the allies win World War II and went on to become writers—people who, through science fiction and fact, let us in on what was ahead. Clarke led the team that developed ground controlled approach (GCA) radar during World War II. In 1945, he published a paper entitled, "Extra-terrestrial Relays," effectively inventing the geostationary communications satellite. Later Clarke told how he was approached by a representative of the Communist Chinese government who planned to launch such satellites to destroy the capitalist West by broadcasting pornography into peoples' homes. Look out, "Sin-e-max."

"What are you most looking forward to in the next 10 years?" he was asked. "Surviving!" he replied. Adding: "Actually—and this is very speculative—I'm 90% sure the energy revolution is starting and the oil age is ending. That is through cold fusion, which is seldom cold and probably not even fusion. The first commercial units are going on the market very shortly."

The authors have a weakness for aphorisms, cliches, quotations, and slogans—in other words, a "bumper-sticker mentality." Before wading into the cold fusion confusion then, a few more quotes from Clarke might be in order:

> *Clarke's first law: when a distinguished but elderly scientist says that something is possible, he is almost certainly right. When he says it is impossible, he is very probably wrong.*
>
> *Clarke's second law: the only way to find the limits of the possible is by going beyond them to the impossible.*
>
> *Clarke's third law: any sufficiently advanced technology is indistinguishable from magic.*

So, what happened to cold fusion?

In 1989, Drs. Martin Fleischmann and Stanley Pons claimed to have produced nuclear reactions by putting palladium electrodes surrounded by coils of platinum wire in cells filled with D_2O and getting excess heat energy out. Clarke professes to believe it. Is he a damned fool? An old fool? A shill? His web site is sponsored by CETI Inc.

In 1995, Clean Energy Technologies, Inc. (CETI) of Texas demonstrated a cold fusion reactor claimed to produce one kW more energy than it consumed. James Patterson, Ph.D. continues to patent palladium and nickel-plated microelectrodes. Whether to avoid the labels of crackpot and fraud, or because he is right, Patterson calls cold fusion, "low energy nuclear reactions" (LENR).

Since the dawn of the Industrial Revolution, rumors have persisted about a tablet that could be dropped into the fuel tank of an automobile to achieve 250 miles per gallon on a tank of water. There have been similar reports of car tires made of a rubber compound that would last for 100,000 miles, or never wear out. Since the U.S. Patent Office opened, so many applications have been filed for "perpetual motion machines" that the office will no longer accept them without a working model—if then.

Of course, discoveries like these would destroy the oil and rubber companies and destabilize the world economy, and so in collusion with the "gummint," patents have been secretly bought up and inventors terrorized into silence—right? Do today's industrial giants really have more capacity to suppress progress than the buggy manufacturers, horse breeders, and whip makers before them? Well, maybe. However, it has been your gentle authors' observation that most conspiracies are about stupidity rather than evil brilliance. Someone does something stupid and then, he, she, or an ever-growing group of "they" are stupid enough to think they can cover it all up. Even if they got away with it, the story will come out.

True—the oil and tire companies would stand to lose if oil were not needed for purposes of heating and plastics, and if tires lasted forever. However, while oil powers cars, trucks, furnaces, and provides the raw materials for the petrochemical industry, less than 6% of electricity comes from oil in the U.S. While more marketing tactic than reality, warranties on car tires now exceed 60,000 miles. It is conceivable to power an automobile with a tablet dropped in a water-filled fuel tank. However, what would be the cost of such a fuel—and what about its environmental impact?

Gasoline prices in the U.S. are always higher than we would like them to be. But while they fluctuate (depending on who is doing the taxing and spending in Washington and the state capitals—not to mention what is going on in the Middle East), fuel prices here are a fraction of what Europeans and people in many other parts of the world pay. Similarly, car tires could be made to last far longer than they do now at the expense of minor performance characteristics like braking and traction.

So, was the sudden demise of cold fusion a conspiracy by the Military-Industrial Complex? Actually, researchers variously describe cold fusion as LENR or, more commercially, "new ^{2}H energy technology," it obviously is alive and well-outside the U.S. Today, Fleischmann and Pons reportedly work in a French laboratory funded by Japanese yen.

The only sure thing about fusion is that despite the ongoing research, "hot" fusion in big-budget reactors has never reached the break-even point energy-wise. "^{2}H bombs" must be triggered by fission bomb detonators, something that would make fusion reactors less desirable. Magnetic containment fields and lasers have been tried as more benevolent triggers, so far without practical success.

So, at worst, cold fusion is as much a pipe dream as hot fusion. At best, it will be an inexpensive kW steam generator in a glass of water. We must also remember the power of market forces. Adam Smith's "invisible hand" only works one way: You can kill an economy but you cannot force it. It is possible to squelch viable economic activity (as the Soviets did with their entire economy!), but you cannot force something that is "uneconomic" to be viable. All the legislation in the world cannot isolate an industry from the liabilities of accidents or pollutants but, wisely mandated, these costs and responsibilities can be included in old-line industries like nuclear generation and cutting-edge future enterprises i.e., fusion energy. Properly run, they can benefit the environment and be cost effective.

Any kid with a battery and a pair of wires can make a saltwater solution in a glass produce bubbles. However, once the academic or energy industry establishment becomes convinced even a micro-J more energy evolves from a cold fusion reaction than must be supplied, the question becomes an exercise for the engineers.

Solar Power Stations

Someone once asked, "How can we be hungry when it's raining soup?"

As we have seen in these chapters, all energy is solar—even if the fuel for nuclear fission does not come from our sun, but from the collapse of other, long dead stars.

If it is raining solar soup, why are we energy hungry? Sooner or later, we are going to use up our coal, natural gas, oil, and uranium reserves. Biomass or crop (corn, etc.), conventional PV, hydroelectric, thermal solar, tidal, and wind power are endlessly renewable, but the total available energy would cause incredible environmental impact long before our appetites were satisfied. In fact, fossil fuels are renewable, too, if we are willing to wait 40 to 100 million years.

Maybe all we need is a bigger ladle with a longer handle. Solar powered satellites, or solar power stations, perhaps.

Solar power satellites have been variously conceived as giant mirrors in space, reflecting sunlight on dark arctic lands such Siberia, or solar collectors generating microwave power beamed to antenna farms on earth. The satellites could be in synchronous orbits, always aimed at the same reception point on the ground or in low earth orbit (LEO), energizing a series of antenna farms under the path of its orbit.

The advantages over terrestrial based renewable energy sources are numerous. Being above the clouds assures weather never would interfere with energy collection. Transmitting power on microwave frequencies not absorbed by water—the opposite of microwave ovens—would minimize weather-related problems of reception of power on earth. Even geosynchronous satellites spend some time eclipsed by the earth, but at 22,500 miles above the equator, they receive sunlight far more than the ground station they serve.

Traditional renewable resources do not add any net heat load to the environment. Like all other forms of fossil fuel and nuclear generation, solar power satellites would. However, except for the final conversion of microwaves back into electricity, all excess heat from inefficiency in the energy collection and conversion process would be dissipated harmlessly in space.

Given the instability of oil supplies in the 1970s, NASA and the

Department of Energy proposed a solar power satellites system (SPSS). The concept required approximately 60 large geostationary PV array satellites. The PV cells, connected as large solar batteries, would have powered microwave oscillator amplifiers connected to precisely steerable narrow beam antennas. Each satellite would have been assigned to a ground station. Because microwave wavelengths are greater than visible light, even transmitting power by a "MASER" (microwave amplification by stimulated emission of radiation—the radio equivalent and forerunner of the LASER)—focusing is not as precise as with light. While the radio wave intensity at the earth's surface would have been only comparable to the energy in sunlight on a hot day, ground stations would have been located in normally unoccupied areas such as deserts or agricultural fields.

If "an elephant is a mouse built to military specifications," getting the elephant back down to flyable weight and dimensions to meet NASA specifications just required adding money—and lots of it. The SPSS was abandoned as too expensive.

However, in February, 1999, despite the chronic state of emergency in their economy, the Russians, with their "just do it," eagerness to try new things and commitment to space research and development, attempted to test a solar mirror satellite, the Znamya 2.5. This satellite would have illuminated various cities around the world for several minutes to demonstrate the potential of solar power from space. Its predecessor, Znamya 2, was observed on the ground in Western Europe and Byelorussia as a 30-second flash of light. Znamya 1 was a prototype never intended to fly. Unfortunately, the experiment malfunctioned and future efforts remain to be announced.

As with any other technology, there are environmental, health, and safety concerns about microwave solar power satellites. Robert A. Heinlein's 1940 novella *Waldo* predicted power being transmitted by microwave coaxial beam—and the impact on the environment and human health. (In the same piece, he also described the remote manipulators used for handling radioactive and toxic materials that came to be called "waldoes.") Arthur C. Clarke published his proposal for geostationary or synchronous communications satellites in 1945. Solar power satellites would combine aspects of both works.

Solar power stations could collect energy using a PV or PV/thermoelectric hybrid array. Peltier, or thermoelectric, devices use semiconductors optimized for generation of electricity when one surface is heated and the other is cooled. Conditions in space are certainly conducive to high temperatures on the sunlit surface of a solar collector and bitter cold in its own shadow. However, just as with terrestrial solar steam turbine generators, large mirrors focused on a central collector promise the greatest efficiency and highest energy output. In the weightlessness of space, aluminized mylar mirrors thinner than a plastic potato chip bag could be aimed at a central collector with minimal support structure.

Solar power from space will probably have to wait until joint public works projects like the international space station give way to real space stations constructed by corporate consortiums. The solar power satellites proposed by NASA and DOE back in the 1970s required more than 20 square miles of PV arrays. Geostationary communication satellite assignments at 1° spacing are highly coveted. These are only 459 miles apart—pretty close by orbital standards.

So far, communications have been the single most profitable aspect of space exploration. Energy from space could be the next real reason to go "out there."

Part 4: Maintenance and Operations and Transmission and Distribution

Chapter 16
Maintenance

Effective maintenance for the many elements of a generating plant fall under three broad classifications:

- normal
- emergency
- preventative

Normal maintenance includes housekeeping items. It can include removing slag and "clinkers" from furnace walls and grates, sludge and scale from boiler tubes by a process known as blowdown and mechanical scrapers, as well as soot from boiler walls and stacks, and soot and cinders from precipitators. It can also include replenishment of feedwater treatment materials, dressing or replacement of brushes on motors and generators, lubrication and replacement of lubricating oils, checking adjustments on governors, regulators, relays and other instruments and arm devices, and routine inspections and checks on other equipment as prescribed either by the manufacturer, the plant maintenance director, or regulatory bodies.

Emergency maintenance includes such items as repair of cracked furnace walls, cracked and broken boiler tubes, broken turbine blades, burned-out or damaged motors, and loose or broken electrical connections. "Emergency situations" are those that mandate the repair or replacement of any element that may effect the plant personnel safety and cause the actual or imminent shutdown of the plant. Any scenario that threatens to seriously reduce the ability of the plant's monitoring devices at the control boards, or the alarm signals to give indications of possible type and location of trouble and procedures, requires trouble-shooting.

Preventative maintenance usually calls for the wholesale replacement of equipment and parts throughout the plant before emergencies strike. These exercises can include replacement of fans and pumps, motors, relays and protective devices, and other critical work such as station lighting, painting, and so forth.

Such maintenance activities for fossil and hydro plants are carried out under so-called "safe procedures" by trained workers using proper tools and equipment under experienced supervision, following a workscope (plan) that has been thoroughly prepared. Where practical, equipment is taken out of service for maintenance work—in some rare cases, it is worked on while remaining in service if the assignment requires it. The skills employed are those that may be found in many other industrial, manufacturing, and commercial undertakings. They include carpenters, mechanics, welders, electricians, test technicians, painters, and other similar crafts.

In nuclear plants, maintenance requirements are generally more stringent as some of the equipment is radioactive, and the risk of exposure to undesirable amounts of radiation is always a factor. In these cases, special equipment is employed to permit handling of radioactive items. Such equipment is often built into the unit as a permanent installation, usually in duplicate (or additional redundancy, in some cases). This also permits some components to be taken out of service without effecting operation of the plant.

Workers in nuclear maintenance jobs are dressed in radioactive-resistant clothes and wear devices that signal the build-up of radioactivity. Their time in the unit is limited and they undergo decontamination procedures at the end of the their shifts. There are many items in a nuclear plant not subject to radiation, of course, and these are maintained in much the same manner as similar items in a fossil fuel plant.

Reliability Centered Maintenance (RCM)

Reliability centered maintenance (RCM) relates to preventative maintenance. It is a term adopted by the Electric Power Research Institute (EPRI) for programs that monitor and predict when certain maintenance may be needed for electric power plants. It's based originally on maintenance programs developed by the airline industry and was introduced into the nuclear power industry as the standard by which preventative maintenance is prescribed for their plants. It has also been adopted by the U.S. Department of Defense.

RCM is designed to enable maintenance work to be planned and carried out in such a coordinated way that system components operate as effectively as possible. It stresses that all maintenance work must be economical in terms of the overall costs and benefits. The program does this by defining and identifying different functions of any given plant system and determining "dominant and critical failure modes" and their causes. It then selects the most cost-effective tasks that prevent most unplanned system outages. Some utilities report to EPRI that they save up to 50% of their previous maintenance budgets by using RCM procedures.

Saving money is the whole point of RCM. Given the increasing regulation that is pulling plant operations in one direction and market forces pulling it in the other, maintenance has become one area of electric power operations where utilities can still achieve substantial cost reductions. RCM is all about applying innovative approaches to maintenance in both nuclear and fossil power plants, as well as power delivery (transmission and distribution systems).

RCM is an essential principle to establish preventative maintenance intervals based on actual equipment performance rather than on a fixed time schedule. Overall, fixed-interval maintenance tends to be too conservative (and thus, expensive) with many routine tasks performed simply on the say-so of the equipment vendor or plant maintenance manager with no regard to how often the component is used or how important it is. On the other hand, essential components can receive insufficient maintenance, resulting in major system failures.

When implemented in all plants a utility owns and operates, RCM techniques can lower routine maintenance costs by reducing unnecessary tasks while improving the reliability and availability of all systems. It does so by focusing on the maintenance needs of the most critical elements only.

This is not to say an effective RCM can be "declared" and implemented overnight and for no cost. The system evaluations required to establish an RCM program involve significant commitments of resources—time, personnel, and cost. However, benefits can be realized almost immediately. In the nuclear power industry, for instance, where RCM was first implemented, the payback period for RCM implementation is now estimated to be only about two years.

Why are RCM-based maintenance programs so much more cost-efficient than conventional maintenance programs-and so quickly?

Plant maintenance teams typically develop conventional preventative maintenance programs a component-by-component basis. They often rely on equipment vendors' recommendations or an attitude that says, "This is the way we've always done things." In contrast, RCM is function-based. It deals not with what the component is, but what it does—and how. It focuses on preventing those failures that are most likely to have serious effects on the overall system functionality.

RCM also seeks to maximize the use of what's called equipment condition monitoring—*i.e.*, taking advantage of what can be learned about a component while it is in service to determine what's wrong when it fails. Such reality checks help minimize downtime or labor-intensive, by-the-book overhauls that gain little. A sound preventative maintenance program based on the principles and practices of RCM can provide improved resource allocation, more useful documentation, and greater reliability of critical systems.

Establishing an RCM program

Several steps that must be followed to initiate an effective RCM program. The first is to conduct a plant-wide failure modes and effects analysis (FMEA) to identify components and equipment essential to each major system function. This is followed by development of a logic tree analysis (LTA) to identify the most effective maintenance tasks that can be used to prevent failure of the critical system components tagged by the FMEA. Finally, a so-called "living RCM program" is established to recognize those maintenance activities that are most effective in reducing failure rates and lowering costs.

A first application of these techniques to power plants not only helped tune up existing maintenance programs and procedures, they had the added benefit of helping to develop new methods and tools that effectively respond to the unique maintenance needs of the power plant environment. Such methods and tools can be shared among all plants to add practice standardization. Maintenance practices had always varied widely among different utilities, with many of them resorting to a strategy called run-to-failure (literally, "fix it when it breaks!"). RCM is expected to substantially reduce these failures and lower total power plant maintenance costs.

Long term, RCM programs have focused on developing an appropriate database of substation equipment types, failure modes, and maintenance tasks. The data are being categorized so that failure modes can be ranked according to their impact on power plant function—*e.g.*, does a failure cause a momentary or sustained outage? How many customers are affected when it happens? Such data will provide essential general information to utilities conducting FMEA and LTA tasks on their own facilities; when shared they can benefit all facilities. Studies are also being conducted to determine the required level of maintenance history reporting needed to track and trend equipment failures for use in RCM "living programs."

Because of the significant differences in maintenance practices among power plants, RCM methods must be validated based on actual experience of utilities participating in the program. Lessons learned from these validation activities can be used to refine RCM analysis for future applications. In addition, maintenance templates can be developed to indicate industry "best practices" for conducting specific maintenance tasks in each type of facility under a variety of operating conditions.

Predictive maintenance

The next step ("beyond RCM," as it were) relies on past experience to set maintenance intervals. This is called predictive maintenance (PM). It uses condition monitoring to give early warnings when specific pieces of equipment are nearing failure.

During the 1990s there were developed new, low-cost sensors enabling so-called "just-in-time maintenance." This made possible even greater cost savings. Some of these sensors have emerged from previously classified military work; others were originally developed for space applications.

Corrective maintenance

Not all failures can be prevented, of course—not even with the most sophisticated monitoring systems.

This means that in addition to preventative maintenance advances, new ways are constantly being developed to lower the cost of corrective maintenance and reduce the time required to make repairs throughout the entire power system.

The Revolution in Power Plant Maintenance

Taken together, these technologies and practices represent nothing less than a revolution in the way utilities maintain their power generating systems.

As mentioned at the outset, many of these and other technologies are already being used in utility power plants. Indeed, the maintenance revolution spawned by development of RCM and advanced sensors is only now beginning to reach into power delivery (transmission and distribution) systems in a major way. EPRI has taken the lead in developing and demonstrating many of these technologies, usually in cooperation with individual utility members and potential vendors.

The future holds even more promise, as a variety of other diagnostic and repair technologies emerge from the laboratory and transform into useful products for utility application. Clearly this is a field of endeavor that benefits the electric power industry as a whole and can have a direct, beneficial impact on consumers. For this reason, it is imperative that the broadest possible industry collaboration is brought to bear on the research and development tasks that remain ahead.

Chapter 17
Operations

Operations is a catch-all term that refers to the coordination of generating plant equipment and processes to produce electrical energy from a fuel source for delivery to end user customers in the residential, commercial, and industrial sectors.

Well trained, highly skilled operators achieve safe, reliable, and efficient production by constantly acquiring and evaluating data, much as a bridge crew uses navigational aids to safely steer their ship. In addition, just as oceans are populated with ships of many different vintages, so the nation's electric grid is energized by plants old and new, each with its own unique modes of operation.

At the time of Edison's Pearl Street Station, in New York—the first generating plant from a century ago—production of electricity was labor intensive. Operations were accomplished manually. Data acquisition amounted to what operators could collect from personnel whose knowledge was limited to their individual interactions with the process.

As technology developed, many plant operations were performed automatically by mechanical or electrically actuated devices. Data acquisition improved, as well—though the number of people needed to operate the plant declined—as more and better information could be gleaned from more complex interactions.

In plants currently being built or where older plants are upgrading their control systems, operations and data control are almost entirely computerized. The personnel required to operate a plant has been further reduced, but operating conditions can be correlated, rapidly and accurately, enabling the fastest possible response to changing conditions. Fewer people on the bridge of the ship today are better able to sail it.

Modern operating and data-acquisition systems include "fail-safe" features and redundancies—duplicate systems or more to back up critical operations—but human supervision remains the final authority over any plant. A century of changes has resulted in higher-than-ever operating efficiencies and lower operating costs to the utility—and to the customer.

Defining the terms

An operations staff oversees the generation of electricity. This can be defined as either gross or net production of electricity.

Gross generation is the amount of power produced by an electric power plant, measured prior to the point at which the power leaves the station and is available to the system. Some of the electric power generated at a power plant is used to operate equipment at the plant. This "in-house use" generally ranges between 1% (for hydroelectric units) and 7% (for steam-electric units).

Net generation is the power available to the system (gross generation less use at the plant). However, it is greater than what is available to consumers due to losses during transmission and distribution (approximately 8-9%).

Those keeping records of electric power production—most notably, the U.S. Energy Information Agency (EIA)—evaluate electric utilities in terms of "net generation" and non-utility facilities in terms of "gross" generation.

Operating and Maintenance Expense: Feeding the Gorilla

Generation—whether measured in gross or net terms—is the 800-pound gorilla of the electric power industry. It accounts for more than half of the average utility's assets and the bulk of a utility's cost of producing and delivering electricity to end-use customers. Its appetite is such that more than a third of the nation's primary energy sources are used to generate electricity.

Let us take a glance at one example of this gorilla's habits.

At this writing, America's 192 investor-owned utilities (IOUs) produce three-quarters of the nation's electricity. (Publicly owned, cooperative, and federal power companies produce the rest.) In 1998, IOUs spent 56% of their operating revenues on expenses (producing electricity and maintaining their plants). Those operating and maintenance (O&M) expenses included such items as:

- wages and benefits
- security
- supervision
- materials and supplies

Fuel for the plants accounted for 32% of all electric plant operating expenses for the IOUs in this same period. This included the costs involved in purchasing, transporting, handling, and preparing the various fuels used to heat the boilers.

Converting those percentages into dollars presents a staggering picture. O&M activities throughout the entire electric power industry—the IOUs and the others—totaled $104 billion in 1998. (This included generation and power purchases, as well as transmission and distribution of the product, customer services, and administrative and other "general activities.") In 1998, power production accounted for almost 45% of O&M expenses, followed by power purchases at nearly 29%, and administrative, general and customer service, and sales expenses at 19%.

It must be added that this expense distribution illustrates, at best, the trend of O&M expenses into the late 1990s. As price competition increases with the entry of so-called non-utility generators—independent power producers and power marketers, among others—it is likely these expenses will trend downward.

In fact, the trends further indicate that IOUs have already lowered some of their historic O&M costs. They have decreased by 22% from about 4.5 cents per kWh in 1986 to 3.5 cents per kWh 10 years later (latest available figures at this writing). Lower fuel prices accounted for most of this. During this same time, non-fuel O&M expenses remained stable at about 2.6 cents per kWh. This is where cost-reduction activities need to be focused.

Those lower fuel prices are primarily due to lower coal prices, which in turn have resulted from excess coal production capacity and changing market conditions. With more than 55% of the electricity generated in the U.S. coming from coal-fired plants, lower coal prices make a big difference in average fuel costs. As coal prices dropped, many utilities found it wise to buy out older, more expensive contracts and increase purchases under newer, less-expensive contracts, or to increase purchases of less expensive coal form the spot market.

Costs for other fuels have decreased as well. Average wellhead prices for natural gas declined between 1987 through 1998. Gas-fired plants produce only about 12% of the electricity in the U.S. but the trend for gas is to take on a greater share. A worldwide surplus of uranium has decreased its prices during the past decade as well. Nuclear plants use enriched uranium to produce about 22% of the electricity in the U.S.

While lOUs have been successfully reducing fuel costs, many have also been able to reduce their workforces and lower payroll expenses. This has been accomplished in the old-fashioned corporate ways—attrition, early retirement, and voluntary and involuntary severance. Between 1986 and 1998, employment at major lOUs dropped by 20%, or more than 100,000 employees. In addition, salaries and wages decreased by 28%, from about 0.7 cents per kWh in 1986 to about 0.5 cents per kWh in 1998. Most industry experts believe that with downward pressures on costs, and increasing automation, staff reductions will continue.

Economies of scale

Samuel Insull worked on Thomas Edison's staff and had a knack for utility operations. As such, he helped to devise economic concepts that still govern modern utility marketing, planning, and pricing.

His laboratory was the Chicago Edison Co., of which he became president in 1892. Chicago Edison, like all electric companies at that time, had high fixed costs (associated with investments in generating plants and transmission equipment) but low operating costs (expenses associated with fuel). He determined that adding customers would "generate" higher revenues and spread out a utility's fixed costs.

In an effort to attract more customers, Insull aggressively marketed the benefits of electricity and reduced its price. The first task was easy enough for any good salesman. However, with no established benchmark price for the commodity at this time, a more difficult task for Insull (and other electric utilities) was establishing an appropriate rate.

What Insull discovered was that the more time a generating plant was in use, the greater its efficiency factor (known as its load diversity). This efficiency yielded higher profits as it lowered costs per kWh. This led him to discover ways of increasing efficiency through economies of scale—using a single plant to service the morning and afternoon streetcar load, the daytime industrial load, and the evening residential load was more economical than using three separate plants, as had been done until then.

On the customer side of the plant, Insull oversaw development of the demand meter, which enabled Chicago Edison to more accurately set a price for the electricity his company sold to consumers. The demand meter did just what the name implies—measured a customer's electric demand (his share of fixed costs required for usage) and the actual energy (or kWh) used. Insull set the price of his electricity to cover both fixed (demand costs) and operating (energy or variable) costs. Fixed costs refer to the fixed amount of investment that must be paid, regardless of output, and include power plant construction and equipment. Operating costs vary with the level of electrical output and include fuel expenses.

So it was that the man who labored at the side of the developer of the practical light bulb also helped nurture the electric utility industry by recognizing the significance of three concepts:

- the inherent advantage in serving numerous customers
- the desirability of load diversity
- the economies of scale realized by building large generating plants to serve customers

Instrumentation

Instruments are the means by which operators keep track of what's going on in their plants. They principally monitor the boiler and the turbine, but all other major mechanical and electrical elements are tracked, as well. Readings are routed to the central control room, where operators can make decisions based in large part on what sensors, and instruments tell them.

What operators read on their instruments (the better term is meters) is data that may be of two types:

- indicating meters that show moment-to-moment conditions
- recording meters that accumulate data over a period of time from equipment or procedures to which they are attached

In the days before computers, all data had to be logged and filed by hand. Today, meters themselves are computerized and their data is electronically true even when the processes involve mechanical quantities (such as temperature, water flow, etc.). Mechanical qualities are converted into electrical quantities by devices called transducers that can be easily transmitted by wires (or fiber optics) instead of piping to the control boards of computer.

"Red Alert, Captain!" Both the indicating and recording meters indicate emergencies via sound and/or light alarms. "Emergencies" can mean abnormal operating conditions up to and including fire alarms anywhere in the plant. Operators hold drills to ensure they can meet emergencies quickly and effectively and to identify areas of their plans that need further work.

Communications. Telephones are used both for internal transmissions and for connection to the outside world. Two-way radios are utilized for internal two-way communication as well. Oftentimes, a plant will have some sort of direct fail-safe link to fire, police, and civil defense departments to ensure emergency situations are quickly communicated to the facility's host community.

Plant Start-up and Shut Down

The most important jobs assigned to any operating group in any plant are the tasks of starting up the various units and shutting them down. These assignments need to be mastered under both routine and emergency conditions. Plants are unique unto themselves but also unto their fuel type and design. It is not a one-size-fits-all world but operators need to adhere to some basic principles.

Steam turbines

Power plant steam turbines work under operating conditions dictated both by the state of the unit equipment and demands of the power system the plant serves. The unit and turbine operating conditions are divided between stationary (or steady state) and transient (or unsteady state).

Stationary operating conditions mean that the power unit operates with constant steam pressure and flow through the turbine producing nominal (rated) amounts of electricity. To reach the maximum electric power load, turbine control valves are fully opened. Everything is "go!"

Because of the sheer size and mass of the turbine rotor, it takes time to bring it up to operating speed. In addition, from a cold start, it takes several hours to generate steam in sufficient quantity and pressure to operate the turbine at all. This relates back to the boiler that brews the steam and to the type of fuel—all these elements affect the rate at which steam is produced, its pressure, and its quality. Some minimum time is required—sometimes days!—to enable the furnace, the boiler (and especially the furnace walls) to expand at a nice even rate and be ready to get to work making steam. In addition, the turbine needs about an hour to "set in" at the new level.

Start-ups are divided into scheduled and "forced" start-ups. The first group includes start-ups from different initial temperature states, changes in the load according to what level of power generation the system requires, and the nature of the scheduled shutdowns for overhauls and inspections. These start-ups and shutdowns are planned and carried out under manual or automated control by the operations personnel. Forced shutdowns are mostly carried out under the purview of automatic systems and emergency warning devices. They are caused by either unexpected changes in the operating conditions of the power system or grid (external) or sudden changes in the state of the unit's equipment (internal).

It is not always possible to draw a distinct boundary between internal and external outages. Occasionally, too, an external or internal outage of a system or piece of equipment may merely mean taking it out of service without bringing down the whole plant. Later, attention can be paid to it during a normally scheduled shutdown and the incident can be charged off to the latter case.

Whether outages are planned or forced, the divisions of labor for them and for start-ups needs to be quite detailed, with all operational personnel assigned a role, which is learned and practiced.

Depending on the duration, start-ups are subdivided into cold, warm, and hot. A cold start-up comes after the plant has been shut down for at least three days. The start-up is considered hot if the shutdown came between 8 and 10 hours previous. All start-ups that fall between 10 and 72 hours are considered warm (or intermediate). They differ in their technical approach; *i.e.*, in a hot start-up, pre-heating of the steam lines is not necessary. These differences in start-up technology depend on the steam generator type and the unit's start-up protocols.

Once a turbine is started, it operates under load (powering the turbine) until it is shut down again, but that load can vary from essentially nothing through "nominal" loads to the unit's maximum rated value. Taken together, this is the unit's governed range. Most of the scheduled changes in a load of a turbine take place within this range. At the lower boundary is the unit's minimum stable load. Moving from this to the governed range—where the real work gets done—depends upon the unit's steam generator and its features (whether its boiler is fossil or nuclear-fueled, for instance). This all requires that operators pay strict attention to the manual and automatic systems by which these changes are made.

At which mode a turbine operates is, once again, governed by internal (plant) capabilities and what the power grid needs from it (or is able to take). That depends on demand—and demand is anything but stable!

In the U.S. and other first-world countries, power generation varies due to the uneven demand during the day and at night, between weekdays and weekends, as well as at different times of the year (heating and cooling seasons). The class of end-use customer also makes a difference (homeowners vs. factories vs. shopping malls, etc.), as does the amount of power shared with other generators or purchased from them. Utilities anticipate these differing needs by maintaining base load and standby ("peak load") plants.

Base load plants form the "base line" of any generating company's output. They are run for as long as possible at full load and designed and built for this assignment by yielding the lowest power-generating costs (and serving the greatest number of customers). The best plants for this class are hydro plants, nuclear power stations, the new class of highly efficient combined-cycle gas plants, and the most sophisticated of the conventional steam turbine plants.

Many utilities also maintain units or entire plants that are used only when demand mandates it (called "peak load" or "peak standby" plants). These plants are uneconomical to operate at all times but can come off the bench and get up and running quickly. Utilities often trade away efficiency to gain flexibility with these units and plants. Gas turbines and jet-engine gas-turbine sets are representative of this group, as are common and pumped storage hydro plants and compressed air-storage plants.

Semi-peak (or "cycling") units provide utilities with flexibility during shutdowns, peak periods during the week, or to assist when other units are started up. Some of them are shut down every night when load lessens and then cranked up the next morning—every day. This is called many things—daily start-stop (DSS); two-shift, or on-off! Many power units and their steam turbines are being specially designed (or retrofitted) for operating in such a cycling manner just as other, larger steam-turbine units are purely base load. In an industry that increasingly has to match production with demand, such fine-tuning is mandatory—even if it means extra work for operators.

Start-ups are always trickier for operating personnel and plant engineers than simple load changes within the governed range of the turbine. If nothing else, there is a higher chance of mistakes when a turbine is started up. Most turbine failures occur either during start-ups or immediately afterwards. This is why operators and plant engineers prefer to keep turbines under load without switching off the generator from the electrical grid. It's also why plants that experience frequent turbine shut-downs (called "trips" if they're unplanned) or who must take them in and out of service to meet demand, modify their unit systems and control boards with automated controls for the most sensitive operations. It also calls for advanced operation personnel training.

When a unit is put on standby, operators try and keep the temperature of the metal in the turbine close to operating range to lessen the range of temperature changes at re-start. This relieves thermal stress on the metal and can shorten start-up duration when turbines (as well as boilers and the other operating systems) are taken out of service for scheduled inspections, repairs, and overhauls. On the other hand, the aim is to cool down the metal as quickly as possible to shorten the length of time the maintenance work has to take. Time is money, after all!

Hydroelectric plants

Aside from their cleaner-than-most environmental record, hydro plants are probably the easiest to start-up and shut down in normal service.

Because the rotor is usually hooked directly to the water wheel and because there is no steam pressure to build, it takes little time to bring the rig up to speed and apply load to the generator. At a shutdown, it is all done in reverse. Load is gradually disconnected from the generator, the water supply is gradually turned off, and the generator and water wheel come to a halt.

The downside occurs in those cases where the electric load is suddenly disconnected. Then the water wheel and generator can accelerate beyond rated speeds and set up stresses on all rotating units.

Nuclear reactors

Start-up and shutdown operations for nuclear turbines and generators are similar to those described above for steam turbines—after all, nukes are steam turbines with power supplied by fission instead of coal or natural gas. Procedures for the nuclear reactor are quite different, however, for the furnace and the boiler.

That is where nukes suffer by comparison, perhaps, with their coal- and gas-fired brethren. Steam in a nuke is generated by heat produced by the radioactive reaction of uranium fuel. The reaction is controlled (heat increased or decreased) by raising or lowering graphite control rods to control the flow of neutrons. Once the reaction is started, resulting radiation makes everything in the reactor chamber permanently radioactive. This means the unit cannot be shut down as quickly and easily as fossil fuel units.

The reactor is shut down in emergency situations by lowering a second set of safety control rods that can be quickly locked down and so halt the flow of neutrons-an action known as "scamming" the reactor.

Chapter 18
Transmission and Distribution: How Generation is Delivered

Even in a deregulated marketplace, delivering electrical energy from the generating plant to business, industrial, and residential customers is not entirely an accounting and marketing effort. Great effort goes into acquiring rights-of-way, engineering design, permitting, building, maintaining, operating, repairing and uprating transmission and distribution systems.

From the switchyard of the generating station to the end-user, electrical energy travels first through the step-up transformer and up to hundreds of miles of wire of the 115, 138, 230, 345 kV (or higher) transmission system.

Figure 18-1 depicts the classic powerhouse-to-your house schematic. The generator output typically is 13.8 kV, seldom more than 20 kV. Generator design engineers must balance the inherent efficiency benefits of higher output voltages versus the limitations of insulation. All generators are really alternators and produce 3-phase ac. The extra wires shown in the figure represent the multiple circuits required to carry the output of several generating units in a large power station. Obviously, any given

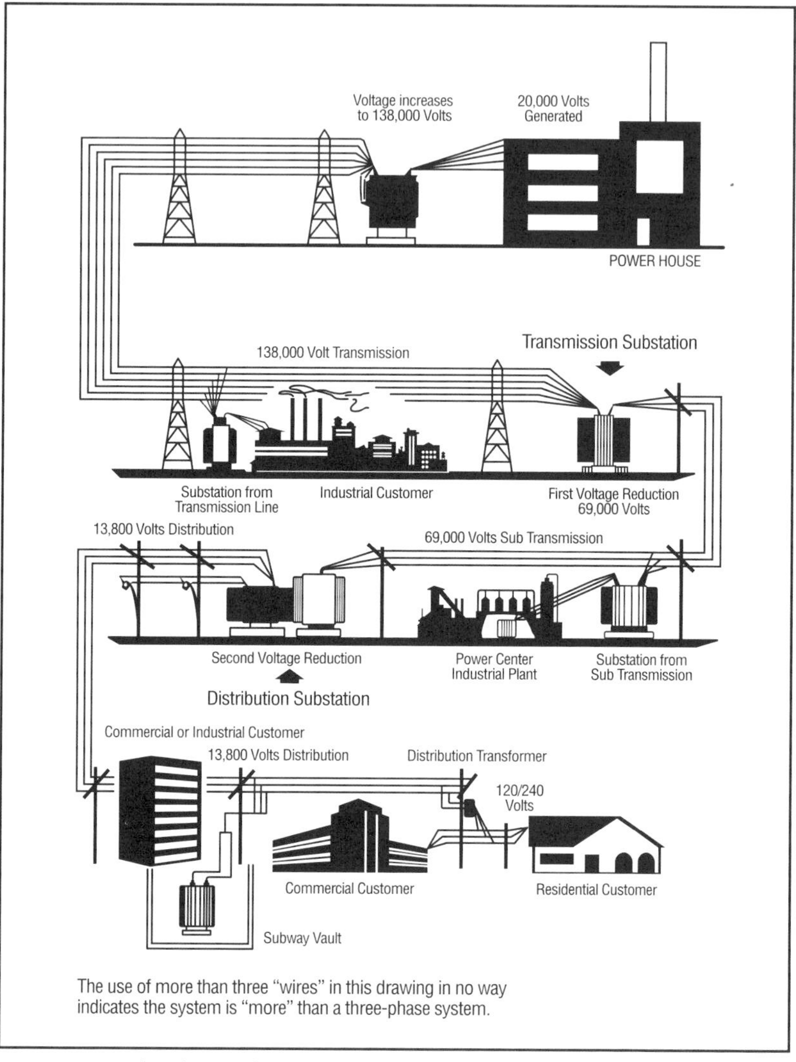

Figure 18-1: The electrical supply system

power plant serves more than one factory, business, and home, but their actual customer base may be a handful of utilities who then serve individual rate payers.

The scheduling aspect of transmission became far more complicated with the beginnings of deregulation in the federal Energy Policy Act of 1992. We glibly speak of "the grid" interconnecting power plants with

Figure 18-2a: End of WMEC Service, Amherst, MA

their customers but unlike the orderliness of lines on graph paper, transmission and distribution lines in the U.S. look more like the nest a yarn factory mouse might build in an electrical control box (Fig. 18-2a, b, c).

Figure 18-2b: End of Mass Electric Service in Belchertown, MA

It is important to remember the technological advance driving de/re-regulation: where it used to take huge fossil or nuclear-fired steam plants to generate electricity economically, the advent of gas turbine combined-cycle generators permit much smaller facilities to generate electricity more cheaply. For nonutilities, who were burning fossil fuels anyway to make steam for heating or manufacturing processes, generating electricity was essentially a "free" bonus! As the benefits of both technology and more open markets were recognized, further rule making by the Federal Energy Regulatory Commission (FERC) and the states allowed nonutilities to sell power to distant customers.

Ac transmission

Transmission lines may be owned by generation or transmission companies, power pools ("grids"), or local utilities. While this has always been true, under deregulation, utilities have been divesting themselves of generating capacity and ownership of the transmission lines may come into question. With good cooperation and planning, the situation will not become as difficult as the telecommunications industry during its transi-

Figure 18-2c: Open 13.8 kV distribution switches

tional years, when pairs of wires within the same cables belonged to AT&T, MCI, Sprint, or other long distance carriers.

The overwhelming majority of transmission lines are 3-phase ac.

Figure 18-3: Transmission lines share rights-of-way with distribution lines

Figure 18-3 shows ac transmission line towers marching up a hill and sharing their right-of-way with distribution poles.

Most industrial and urban areas developed near bodies of water—lakes, oceans, and rivers—that made transportation of fuels such as coal practical and provided water for steam turbine generators. Rivers also offer hydroelectric power opportunities. Not surprisingly, most electric generating plants are located adjacent to water and as near their customers as possible. However, natural resources such as those of Hydro-Quebec's may provide energy from consumers long distances away.

Transmission lines and structures

Electric energy may be transmitted overhead, underground, even underwater. Poles or towers support overhead conductors. Underground cables may be armored or jacketed for direct burial or their insulation protected in conduits or structures designed for multiple cables. A submarine cable may lie on the floor of the body of water or be buried by the cable laying vessel in a trench dug for it. As appropriate, submarine cables must

Figure 18-4: H-type transmission structure

be protected against boat anchors, dredging operations, fish nets-even sharks, which are able to detect the electric fields surrounding the cables and are likely to take a taste.

Figure 18-4 shows a 345 kV transmission line connecting the Vermont

Yankee nuclear power plant and Northfield Mountain Pumped Storage Station with the ISO New England substation in Ludlow, Mass. The tall H-style towers provide plenty of clearance over brush that may grow up along the right-of-way, distribution lines, railroad tracks (foreground), and roads. Note the doubling of conductors on this single circuit transmission line. Double conductors help overcome both the "skin" effect, which causes current on ac lines to flow mostly on the outer surface, and to reduce corona. Transmission lines use suspension type insulators as shown. The top wires along the towers are the multi-grounded neutral and grounding conductors.

High voltage gradients around conductors or on the surfaces of insulators strip air molecules of their electrons. This ionization—called corona from the Latin word for crown—makes a faint buzz, especially on humid days. Only rarely is the glow for which corona is named visible. Corona wastes some electric energy and accelerates aging of insulators—especially ones made of organic materials—and may lead to destructive arcing to poles or towers. Corona also emits radio waves, causing interference to AM radio receivers. While it cannot be entirely eliminated, corona may easily be minimized by adherence to good construction and design practices. Large-radius corona rings surrounding connecting bolts and sharp wire ends limit large voltage gradients to prevent corona within their sphere of influence.

Dc transmission

Thomas Edison got out of the generation business after losing an emotional battle against the perceived dangers of ac. He and others actually feared the high transmission and distribution voltages that made ac so efficient, and therefore economical, rather than ac itself. Edison's illuminating plants distributed dc power at the 110/220 V level used by incandescent lighting circuits in homes and businesses. Unfortunately, at such a low voltage, the illuminating plant could be no more than two miles from a customer. Ironically, high tech transmission lines now use the dc Edison favored.

However, among Edison's contemporaries only Nikola Tesla would have felt comfortable with the UHVs employed. Tesla developed 3-phase ac power systems and invented the brushless induction motor used widely today that runs on ac. Without the development of 3-phase transmission and distribution, Edison's preference for dc might well have won out. Tesla also experimented with very high voltages. The SI metric system unit for magnetic field flux density is the Tesla (T).

Edison's name is inextricably linked with industrial giant GE. Interestingly, with his well-designed dc distribution systems, dynamo, and incandescent lamp, he was interested in electricity only for illuminating purposes. Bitterly opposed to ac, Edison had to be content to cede its development to his competitors and partners, including Samuel Insull (a financier who began his career as a secretary to Edison and rose to notoriety as president of the Chicago Edison Company. He did not invent insulation). Stanley, Steinmetz, Tesla, Thomson, Westinghouse and others used Edison's name to develop the industrial applications of electricity and paid royalties for using his patents.

From the alternator in your car to the largest nuclear steam plant, today's generators produce 3-phase ac that provides the same smooth transfer of power as dc from a battery, but is easier to manage. (Diodes in the car's alternator immediately rectify the ac to be compatible with the car's dc electrical system.) A transformer in the switchyard of the generating plant increases the 13.8 or 20 kV output of the generator to high voltage, extra high voltage (EHV), or ultra high voltage (UHV). EHV is greater than 230 kV and UHV is greater than 800 kV. For dc transmission, high-voltage rectifiers convert the ac (EHV or UHV power) to transmit dc using a positive conductor, grounded neutral, and negative conductor scheme.

A dc transmission system requires only two energized conductors, a grounded neutral, and a grounding shield, while 3-phase ac transmission needs three live conductors plus a grounded neutral and a grounding shield. At the other end of these special-purpose dc transmission lines, electronic inverters convert the energy back to 3-phase ac for connection to conventional transmission lines.

Mile after mile, the benefits provided by these UHV dc lines really add up. While every technological advance has some tradeoffs, dc transmission requires only two conductors of the same gauge per circuit versus three for ac. Note that both systems require multi-grounded neutral (MGN) and grounding conductors.

The nominal phase-to-phase voltage (138, 230, 345, 500 kV, etc.)

describes ac transmission lines. Divide the nominal voltages by the square root of 3 ($\sqrt{3}$ = 1.732) to find the phase-to-ground voltages of 79.7, 133, 199 and 289 kV, respectively. Dc transmission lines are designated by their conductor-to-ground voltage; *e.g.*, a ±450 kV line would be called 900 kV if it were ac!

Because the voltages are usually higher in dc transmission, insulators are longer and towers larger to accommodate greater clearances. However, ac insulators must withstand the peak line-to-ground voltage and the constantly changing electrical fields stress insulators more than dc, so the differences are not as great as you might expect. Multiply the phase-to-ground voltage by the square root of 2 ($\sqrt{2}$ = 1.414) for the peak insulator voltage. Therefore, on a ±450 kV dc line the insulators must withstand 450 kV between conductor and ground—only about 10% more than the 408 kV peak on a 500 kV ac transmission line—but all other factors being equal, the ±450 kV dc line carries about four times the power!

Rectifying the ac at the transmitting end and inverting the dc back to ac at the receiving end adds two extra steps not required in conventional ac transmission systems. Because dc lines are normally long haul, the inverters at the terminating end also perform the phase-shifting function that would be necessary to synchronize a comparable ac line with the local grid.

Higher voltages mean significantly more power can be transmitted using the same sized conductor. Doubling the voltage means transmitting four times the power and tripling yields a gain of nine times. With efficiencies already greater than 99.5%, this means more energy can be transmitted longer distances using fewer circuits. This saves the cost of acquiring new rights of way and building more transmission lines.

Caveat nominator

Remember that nominal transmission and distribution voltages vary by region and utility. This book will continue to refer to 138 kV as the "typical" transmission voltage, even though it may never have been used in your area, and with uprated lines becoming more common, that may be low by a wide margin. (Do not be fooled by Mercator projection maps: Greenland is about a quarter the size of Australia, and at 71,997 square miles, all of New England, including Maine, is only slightly larger than

Missouri or Oklahoma. With such a small region to cover and with an older infrastructure, New England gets by nicely with predominantly 115 kV transmission. Multiply the 115 kV by three and its EHV lines operate at 345 kV. Of course, it has ±450 kV dc lines bringing energy from Hydro Quebec, also.)

The voltages really are only "nominal." A 345 kV line operating at the top end of a 5% limit is actually 17.25 kV high, or 362 kV. This difference is 150 times the "nominal" standard multiple of 115 V used to specify transformers.

Subtransmission

A transmission line terminates at a substation, where the transmission voltage is "transformed" to the subtransmission level—nominally (!) 69,000 V. Subtransmission lines belong to the local utility and radiate to serve distribution substations. At each distribution substation, a transformer steps the 69 kV voltage down to the typical distribution voltage of 13,800 V (13.8 kV). Pole-mounted transformers further reduce the voltage to 120/240 for use in homes and small businesses.

Efficiency

Ohm's laws rule efficiency. Just as in the generating station, every effort must be made to insure efficiency. The longer the lines and the heavier the loads, the more important efficiency becomes. Capacitors and transformers in transmission and distribution substations can compensate for voltage drop, but once the energy is wasted as heat, it cannot be recovered. Remember, the power wasted as heat depends only on the square of the current running through a conductor and its resistance. Therefore a stretch of line carrying 600 A with a resistance of 1 Ohm dissipates 360,000 W! If this is a 138 kV transmission line, the circuit transmits 82.8 MW of power (less the heat loss) for an efficiency of 99.57%! However, for a 13.8 kV distribution line at the same 600 A and 1 Ohm values, the power carried is 8.28 MW; while the loss would be the same, the efficiency is reduced by a factor of 10 to 95.65%. While a four-point loss hardly seems significant, losses in transformers, transmission lines, and distribution networks multiply every step of the way until minor losses become significant:

(.9957) x (.9957) x (.9957) = 0.9872 0.9872 x 100 = 98.72 % Efficiency

(.9565) x (.9565) x (.9565) = 0.8751 0.8751 x 100 = 87.51 % Efficiency

Clearly, an 11-point difference in efficiency would not be tolerable.

When a connection, switch, or transformer begins to deteriorate, initially the current through it does not decrease appreciably because its resistance is still a small fraction of the total circuit. However, the power dissipated in the defective component increases linearly with the resistance, sometimes yielding spectacular results. Outages may occur, at worst injuring or killing bystanders or personnel and, at minimum, inconveniencing ratepayers, reducing revenues and increasing costs for the affected organization.

Voltage drop and power loss

The various aspects of Ohm's laws were discussed in chapter 1. Resistance and impedance in any conductor cause voltage drop and power loss. Just as any pipe obstructs the flow of water to some degree, any conductor resists electrical current flow. An iron wire resists current more than an aluminum wire of the same diameter and in turn, the aluminum exhibits more resistance than copper wire of the same size.

Because of persistent safety problems with wiring devices and installation methods, aluminum wiring is rarely used in household applications. Unless protected with an antioxidant paste and scoured to remove the naturally occurring oxide coating that makes aluminum so durable, an aluminum connection has a relatively high resistance that only gets worse when heated, further oxidizing the joint and increasing the resistance. Also, aluminum is more of a "super cooled liquid" than a solid metal. Unless specially designed clamps, connectors, or screws are used, the aluminum "cold flows" under pressure, inevitably loosening the connection, increasing the resistance, or even causing arcing. In residential wiring, aluminum wire must be two gauge sizes larger than copper to be rated for the same current. A 14-gauge copper Romex cable carries a 15 A circuit, but with aluminum wiring, the same circuit requires 12-gauge conductors. To handle a 20 A circuit safely, 12-gauge copper or 10-gauge aluminum wires are required.

However, the 600 A transmission or distribution line described above is probably an "aluminum clad steel reinforced" (ACSR) conductor. Tradeoffs in conductivity, strength and weight require innovative engineering. All copper conductors initially would provide the lowest resistive losses, but poles and towers would have to be close together. Aluminum and copper are also highly ductile and inadequately supported 4/0 conductors would soon be magnet wire trailing on the ground. Steel wire is strong but a relatively poor conductor. Because of the "skin effect," the majority of ac current is conducted on or near the surface, allowing aluminum cladding to conduct most of the current while the steel core carries the mechanical load.

Voltage drop is the electromotive force (EMF) difference (V2-V1) that develops between any two points on a current-carrying conductor such as a transmission or distribution line. The voltage drop along miles of line is impractical to measure the way you might check the voltage from plug to outlet of an extension cord used to power a tool. Voltage drop is sometimes called "voltage rise" because, while resistance in the conductor reduces the voltage available to the intended load, the voltage measured across two points along the conductor would be greater than zero and proportional to the current and resistance of the wire.

For example, we used a digital multimeter to measure the voltage drop from a duplex receptacle to the end of a light-duty extension cord while operating a circular saw plugged into it. The 13 A rated extension cord measured 0.4 V from end to end on both the "hot" and "neutral" wires while running the 10 A circular saw unloaded for a total voltage drop of 0.8 V. (*Warning*: Do not try this at home. We are professionals!)

Ohm's law can also be used to calculate voltage drop if the current, the length of line, and the resistance per unit of length are known. A transmission line is 48 miles from the step up transformer in the switchyard of the generating station to the first transmission substation. The 1,000 MCM gauge ACSR conductors have 0.1046 Ohms resistance per mile. (Note that wire tables in reference and textbooks typically list wire resistances in Ohms per 1,000 feet.)

(18.1) $$R = (\text{miles}) \times \frac{Ohms}{mile} = 48\ \text{mi} \times \frac{0.1046\,Ohms}{mile} = 5\ \text{Ohms}$$

(18.2) $$E = IR = (600\ \text{A}) \times (5\ \text{Ohms}) = 3000\ \text{V drop}$$

where:
E is electromotive force in Volts (V),
I is current in Amperes (A) and
R is resistance in Ohms.

One of the most interesting challenges in researching this book was trying to understand the efficiency of transmission and distribution systems. If electric heat is 100% efficient—and it is—why is electricity the most expensive way to heat a home? Modern fuel oil, natural gas, and propane furnaces are "only" 97% efficient. Even if the mark-up and taxes were comparable, surely utilities must get a better deal on fuel than homeowners can from their local heating oil dealer or natural gas utility! So why does it cost so much to heat with electricity?

A 3,000 W electric baseboard heater operating on 240 V ac has a resistance of 19.2 Ohms and uses 12.5 A of current. If the transmission and distribution systems effectively added only 1 Ohm to the heating circuit, about 5% of the heat you are paying for would be unavoidably wasted in the lines and transformers leading to your house. Heating electrically would be only 5% more expensive than oil or natural gas, and the cleanliness, convenience, quiet, and safety of electric heat would be worth the premium. Unfortunately, with the resistance of an electric heater so low, a large portion of the heating element is strung for miles on poles between the generator and your house. Electricity is convenient and economical for so many applications. Heating is not one of them.

Not to lie to you, but there are two other factors that effect efficiency: corona and reactance.

Corona

High voltage causes corona when electric field gradients develop unequally across insulators. In such cases, sharp edges on energized hardware or tight radius bends on conductors exceed the insulating capabilities of the surrounding air. Under ideal conditions, corona is visible as ionized air. When the air is humid, corona can be heard as a crisp buzzing like bacon frying. To prevent corona, insulators are made larger than would be necessary just for straight voltage standoff and creep distance. Large diameter metal corona rings surround the hardware connecting conductors. Corona is more of a problem at higher altitudes. Ironically, while both dry, sea-level pressure air and vacuum make excellent insulators, low-pressure air does not. Not surprisingly, because increasing humidity reduces air density and the moisture may combine with airborne pollutants; high humidity also makes air a poor insulator, increasing corona.

The sharp points on lightning rods discharge static electricity in the air near them, actually making a lightning strike less likely on a protected building. Conversely, the little ball on the end of an automotive radio antenna is not there to keep you from poking your eye out when washing the car. Along with the rubber tires insulating the vehicle from the ground, this corona ball effectively prevents the antenna from being struck by lightning. With its steel frame and sheet metal forming what is known as a "Faraday shield," lightning striking the car is the least of your worries during a thunderstorm. While corona discharge wastes some energy, the benefits of higher voltage transmission lines outweigh the costs of designing and uprating them.

Reactance

Inductive reactance acts as a load. It causes transmission lines, transformers, and motors to use more electric current than is actually consumed in converting energy. However, these circulating currents require generation, transmission, and distribution systems to produce and carry more electricity than is required to perform normal energy conversion or usage functions. The heat loss in the lines is real even if the "reactive power" in transformers and motors is not. The total "real power" plus reactive power is measured in V-A, kilovolt-ampere (kVA) or mega volt-ampere (MVA), depending on the magnitude. Similarly, V-A Reactive, kVAR, and mega volt amperes reactive (MVAR) describe the reactive aspect only. Capacitive reactance appears as a voltage source and can be used to correct for inductive reactance, bringing the system close to "pure" resis-

Figure 18-5: Capacitor banks on a distribution pole

tive W, kW, and MW. The rectangular banks of capacitors often seen on utility poles perform this function as discrete and moderately expensive components (Fig. 18-5).

Fluorescent lighting is inherently "capacitively reactive." It is not only preferable to incandescent lighting because it saves energy but also partially corrects for the inductive power factor of industries and businesses with motor-driven equipment.

Underground transmission and distribution

Electrical transmission and distribution lines are traditionally built above ground. In much new construction, such as new housing developments, underground distribution is worth the extra cost. In the central business districts of major cities, both transmission and distribution lines must be underground with other utilities. Various schemes for insulating underground lines were used from the earliest days of electrical and telegraphy lines in cities.

Transmission and distribution lines are entirely separate. (Look at Fig. 3-1 again.) They connect only at distribution substations, although in heavily developed areas the wooden poles of distribution systems may march along rights-of-way with much taller transmission structures—often, with different owners. Large towers of various geometries carry the 138,000 V and higher transmission lines across the countryside through fire breaks, along highways, or railroad lines. Distribution lines follow secondary roads and streets. Utility poles carry the 13.8 kV high-voltage distribution wires, step down transformers, 120/240 V wires to houses and businesses, and CATV and telephone cables. Some large industries have their own substations connected directly to transmission lines. Most electric customers, however, receive their electric energy from the distribution network of the local utility.

The 13.8 kV distribution voltage lines run along opposite ends of the highest crossbar of utility poles. Traditionally, the wire daisy-chained from pole top to pole top was the ground wire. Having the ground wire at the highest point provided protection against lightning. However, with an additional insulator and liberal use of fused cutouts and lightning arrestors, the former ground wire position may be used for the third phase conductor, conserving materials and increasing efficiency. Three-phase systems provide continuous, smooth energy transfer—not just smoother than an 8-cylinder engine, but inherently smooth. Single-phase power is interrupted 120 times a second, comparable to a 1-cylinder lawn mower engine. The "multi-grounded neutral" wire mounted lower on the pole (but above other utilities) physically supports 120/240 V distribution wrapped around it. In rural installations where single-phase distribution branches off from 3-phase, the grounded neutral conductor provides a return for the 13.8 kV primary winding of pole transformers (Fig. 18-6).

At higher overall cost, using taller poles eliminates the need for a crossbar and ground wire, and 3-phase conductors attach one above the other directly on the poles. Pole mounted transformers reduce the 13.8 kV to 120/240 V for residential customers or 480/277/120 V or 208/120 V for business and industrial users.

Note that while even engineers variously describe the basic main supply voltage as 110, 115, 117, 120 or even 125 V, the voltage standard for the U.S. is 115 V. Transmission and distribution transformers are rated in multiples of 115 V. However, 120/240 single-phase or 120/208 V 3-phase distribution is actually now more common. Several factors account for most of the confusion:

- Thomas Edison's illumination dynamos distributed 110/220 V dc
- 115 V is a nominal value with a tolerance of 5 or 10%

From lamps to motors, most electrical devices operate more efficiently on higher rather than lower voltages. Therefore, secondary distribution aims at the high side of the limit. One hundred twenty V is within the 5% upper limit of 120.75 V and has become the de facto standard. Circuit conductors within a building must be rated to meet the National Electrical Code (NEC) and prevent voltage drop below the 5% lower limit of about 110 V.

Industries with 208 V 3-phase service inherently have 120 V available for lighting and standard appliances such as coffee makers, computers, copiers, etc. The 3-phase voltage divided by the square root of 3 determines the single-phase line-to-neutral voltage. $208V \div \sqrt{3} = 120.09$ reinforces the de facto standard of 120 V.

Businesses and industries served by 480 V have the advantage of being able to operate fluorescent lighting on 277 V. This highlights one of the benefits of ac: transformers, whether they are single or 3-phase, may be connected differently on the secondary side than they are on the primary. This provides many possible voltage and current combinations. (Fig. 18-6)

For the purposes of this volume on electric power generation, most

Figure 18-6: Three-phase distribution

principles are similar for the transmission and distribution of electric energy, except voltage levels tend to be much higher on the transmission side. Cost effective and efficient transmission and distribution involves factors having nothing to do with electricity or Ohm's law, but are economic, political or purely practical, such as the design of structures to meet the challenge of extreme weather conditions. Economics effect where rights-of-way for new transmission lines are obtained and politics help determine the color and geometry of the structures.

Load capacity, including sizing for power factor and peak demand, is the main technical issue. Generating capacity must meet or exceed actual delivered energy, resistive losses in transmission and distribution, and apparent (reactive) MVA load over MW load because of power factor and peak demand. Transmission and distribution networks face similar requirements.

The most important benefit of the "multigrounded neutral" system used in the U.S. is safety. Under normal conditions, the return currents from all conductors balance and cancel each other so little current flows through the neutral conductor and none to or from the ground connection. While the earth as a whole is a nearly perfect conductor capable of conducting any amount of current, interfacing with the ground at any given point is less than ideal. Even a full 8-foot ground rod in good soil may exhibit about 100 (Ohms resistance. Routinely running even a small current through this interface would lead to corrosion and heating with contamination of the soil by the electrode. Adding salts such as copper sulfate to the soil to improve soil conductivity is a marginal answer in these environmentally conscious times. Multiple ground rods help, but in 120/240 single-phase, 3-phase ac, and high voltage dc systems, the neutral carries any current imbalance, leaving the grounding conductor and ground rods in good condition to handle a fault, a lightning strike, or system-created surges and transients.

Anatomy and physiology of transmission and distribution lines

Because transmission and distribution lines overlap in current, structure and voltage, both will be described here.

Structures. Overhead transmission and distribution structures are classified as poles or towers. Either may be made of metal, reinforced concrete, or wood. Wooden poles and towers must be set below the frost line and deep enough to prevent movement from normal loading and weather related stresses such as icing and wind. As a rule of thumb, crews set poles 10% of overall length plus the thickness of the pole at the bottom end below grade in firm soil. Reinforced concrete footings anchor metal poles and towers.

Guy wires. Poles or "A" and "H" towers carrying tangent (straight) lines with equal loading need not be guyed. "V" and "Y" towers must always be guyed. Where the transmission line changes direction or has unequal line loads—as at a "dead end" where a transmission line feeds a substation, or goes underground—the structure must be guyed. Good design and construction practices allow small angles (up to 10) to be guyed with a single guy splitting the angle. Sharper angles must be supported with a guy opposing each conductor.

Look again at Figure 18-4—the 345 kV "H" tower. Note that even though the conductor angle is low, two sets of guys keep the structure of this transmission line stable.

Figure 18-7: Rural 3-phase distribution benefits both utility and house-turned-restaurant.

The dead end in Figure 18-7 adapts overhead 13.8 kV 3-phase to insulated cable in a polyvinyl chloride (PVC) conduit that runs down the pole and underground to a commercial building housing several shops and a restaurant serving the best breakfast in Amherst, MA. The guys brace the pole against the static load of the cables and conductors, as well as the changing dynamic loads caused by ice, snow and/or wind loads.

Figure 18-8 depicts the guying arrangement of distribution poles supporting lines at right angles to each other at a busy commercial corner.

Installing a guy wire at each conductor level on poles reduces shearing forces. Distribution poles typically support 13.8kV single or 3-phase conductors on crossbars at the top. Below the 13.8 kV lines, 120/240 distribution wraps spirally around the multi-grounded neutral conductor. Originally the lowest voltage on the pole, it was logical to install telephone cables closest to the ground. As a relative Johnny-come-lately, cable television cables got stuck between the telephone and the 120/240 V distribution lines. In practice, on distribution poles, guy wires are installed at the level of the 13.8 kV distribution lines and telephone cables.

Where a transmission or distribution line follows the inside curve of

Figure 18-8: Three single-phase transformers efficiently distribute 3-phase energy to rural restaurant.

a road, a horizontal cable may transfer the load across the highway to another pole anchored by a guy wire. Where guy wires and stations cannot be built outside the angle of the transmission line, a pole attached to the structure at the level of the transmission line to provide counter pressure may be used (Fig. 18-9).

Insulators. Mushroom-like "pin" type insulators supported on crossbars from below may be used on distribution and subtransmission lines up to 69 kV. Above 69 kV, strings of suspension insulators tailored to the line voltage solve problems.

Several factors affect insulator design and rating. To begin with, two adjacent conductors in dry air at sea level must be about an inch apart for every 75 kV peak voltage. A 138 kV ac line has about 195 kV peak voltage for an initial distance of 2.6 inches. As humidity increases, so must the gap. Unfortunately, unless you live in Death Valley, you will rarely find dry air at sea level. Even in seacoast cities like Galveston, San Francisco, or Bar Harbor, the topography does not remain "sea level" for long, again requiring the gap to widen. Rain complicates the issue even further. While very pure water insulates well at high voltages, a tiny solid speck forms the con-

Figure 18-9: Former power plant now houses community-access TV studios

densation nucleus of every raindrop. Rain washes pollution (natural and otherwise) from the sky, reaching the ground with pollutants from pollen to smoke particles to vehicle exhaust to volcanic ash. Typically it has a highly acid pH of 4.5, indicating the presence of hydrogen (H^+) ions and carbonate (CO_3^-), nitrate (NO_3^-) or sulfate (SO_4^-) radicals. Organic contaminants accumulating on the surfaces of insulators provide breakdown paths, reducing insulator effectiveness. The acids etch the glazed surface of the ceramic insulators, weakening them and allowing more contaminants to accumulate. So, just as "an elephant is a mouse built to military specifications," the size and number of insulators is much greater than would be necessary under elusive ideal conditions.

Basic impulse insulation level (BIL). Just as adrenaline allows you to perform heroic feats of strength in an emergency, which even intensive muscle training would not allow you to do continuously, insulators can withstand very brief voltage spikes during lightning or other surges. Such surges would destroy them if those voltages were applied continuously.

Organizations representing electrical equipment manufacturers have developed standards and testing methods to be able to specify the BIL rating. The BIL test runs more than 50 microseconds (s). In the first 1.2 s, the applied voltage rises from zero to its peak and after a total of 50 s decreases to half its peak voltage. Under these conditions, the highest peak voltage the equipment withstands without arcing is its BIL rating. Typically, the BIL rating is about 1.5 times the continuous or repetitive peak voltage rating.

Conductors. Only platinum conducts electrical current better than silver. Among conventional materials, copper comes very close to silver. Gold beats aluminum by a smaller margin than it is beaten by copper. Some day "high temperature" (not cryogenically cold) superconductors may carry electric energy. Until then, aluminum, copper and steel, in various combinations, will do the job.

With high conductivity and low cost, copper is the best choice where its low strength and high ductility are not a problem. Where strength is required, copper clad aluminum or ACSR conductors must be used. These conductors must be oversized relative to copper because of their higher resistivity (Table 18-1).

Relative Resistivity	Material	Comments
7	Iron (Fe)	High strength, heavy. Moderate conductor
6	Nickel (Ni)	Conductive oxide protects connections. Moderate conductor
5	Aluminum (Al)	Strong, light. Good conductor. Nonconductive oxide, softness require special handling
4	Gold (Au)	Precious metal highly resistant to oxidation protects connections. Very good conductor
3	Copper (Cu)	Low cost, excellent conductor
2	Silver (Ag)	Semi-precious metal. Excellent conductor, low oxidation
1	Platinum (Pt)	Precious metal. Excellent conductor, low oxidation

Table 18-1: Conductor Resistivity

Underground transmission

Project managers building underground transmission lines must be prepared to hire the really hard-working crews who are not satisfied to leave 40 or 50 feet of pole sticking out of the ground.

What could be better than underground transmission? It is safe from automobile accidents, lightning, vandalism and weather, but unfortunately, underground installations cost more every step of the way.

Circuits

The circuit breaker panel in a home distributes electrical power throughout the house using two dozen or more circuits. Dividing the load allows circuit breaker ratings to be low enough to protect the appliances, cooling and heating system, lamps, and other loads they serve, as well as protecting wiring from overheating and causing a fire. When work needs to be performed on one part of the electrical system, just one circuit breaker may be switched off, allowing life to continue normally in the rest of the house.

Similarly, even when a transmission line effectively serves only one utility, the load may be split among two or more circuits. By adding a second set of three conductors and an additional grounding shield to the towers, capacity is doubled. In a growing region where demand is increasing dramatically, if the transmission voltage is increased at the same time (normally by a factor of three) the two uprated circuits carry 18 times the capacity of the original transmission line! In the real world, many factors affect plans to uprate a transmission line so capacity may not increase so dramatically.

Grid, what grid? A memoir

Figure 18-10 is a grid. A power pool—or what is commonly called "the grid"—does not look like this. What we do have is troublesome at best!

The authors grew up in Pennsylvania about 30 miles north of Philadelphia. In 1965, all of us tended to listen to New York radio stations, which came in loud and clear in southeastern Pennsylvania. Barnett's father was city editor of the local newspaper and for some reason, the day of the Great Northeast Blackout, he was driving Dave around on his paper route. While the lights only flickered briefly, they listened intently as

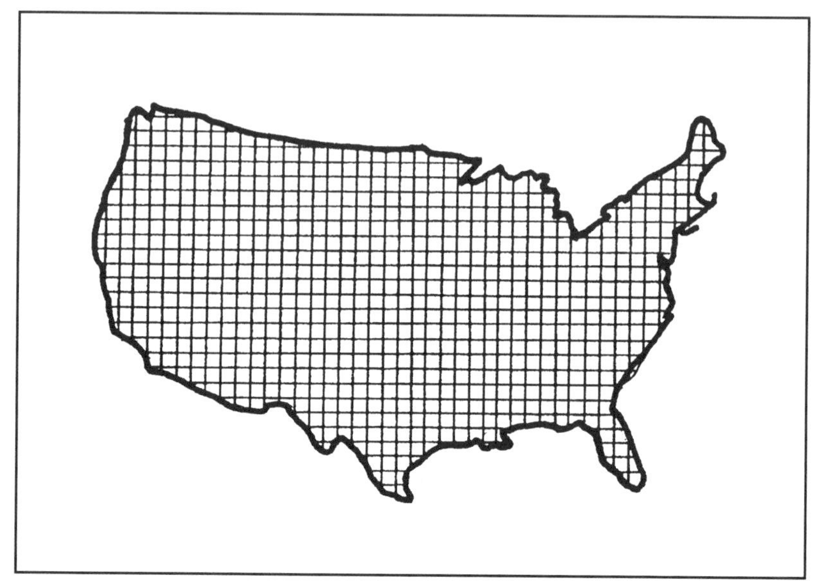

Figure 18-10: A grid.

reporters on the leading stations of the day—WOR, WABC and WCBS—told the story of a region without power. They were broadcasting from darkened studios while emergency generators in the high-rise buildings kept the microphones alive. At their transmitting sites across the Hudson River in New Jersey, more back-up generators proclaimed civilization was down, but not out.

Through the evening they listened as the drama unfolded. Early optimistic pronouncements by governmental and utility spokesmen gave way to more realistic assessments of when power would be restored. As a budding engineer and journalist, Dave took notes as he listened to the radio in his own darkened bedroom. Those notes are reproduced here unedited.

Figure 18-11 is a United Press International (UPI) wirephoto his father brought home the next day showing the extent of the blackout. Remember, both the reports and Dave's notes are from 1965, when the "Big Apple" was still the center of the world as we knew it.

"Tuesday, November 9, 1965: Power Blackout
Hartford and Waterbury were first to regain power.
Con Edison predicted lights would be on again by 7:55.

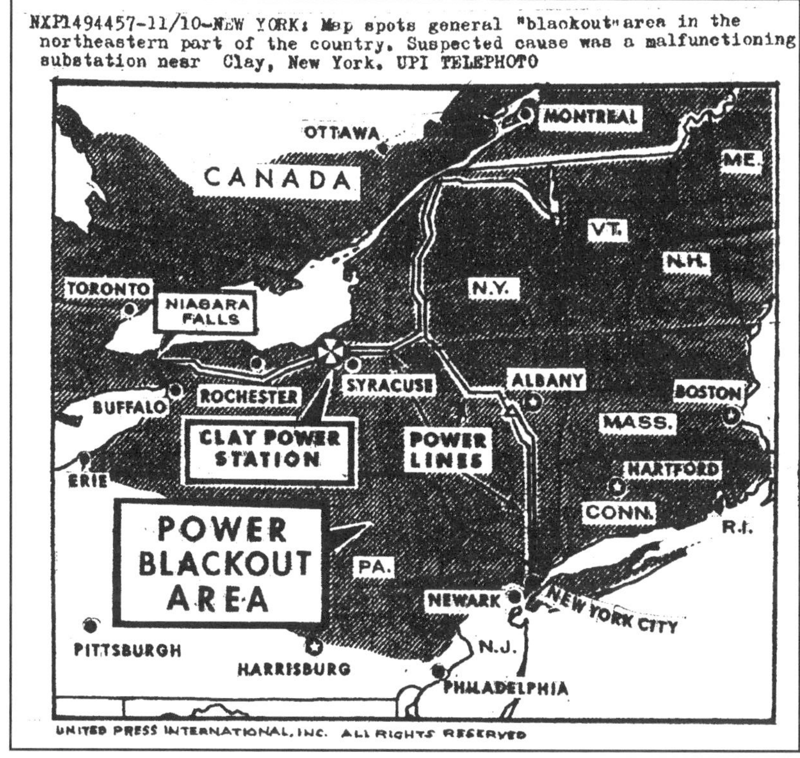

Figure 18-11: "The Great Northeast Blackout of 1965"

Middle Section New Jersey and various areas in the city of New York. Southwest Brooklyn has lights now.

Much confusion in New York but no panic at 7:30. Newark Airport operating normally.

Con Edison, Boston Edison, New York Authority and Niagara Mohawk were all affected.

Niagara Mohawk said a power tower that was down causing the failure. Earlier an unidentified ham had reported this.

An official report from Washington said at 10:30 pm that power would not be restored until 3:00 am.

Looting started in Uptown Manhattan and Brooklyn. Some looters were arrested in Brooklyn.

The fire departments complained that there was a tremendous outbreak of small fires all around New York. In Queens where water pressure has failed the new super pumper is being used to fight the fires.

10:45 Power was coming on in upper and lower Manhattan and Williamsburg section of Brooklyn.

In various jails in the affected areas prisoners took advantage of the blackout to try to escape.

11:00 Power was steadily going on on Long Island and Manhattan.

Gasoline was unobtainable in the affected areas because the gas pumps are electrically run.

The governor of New York, Nelson Rockefeller, said that the blackout is plausibly sabotage.

Failure was most of New York State and New England, and spread to Harrisburg in Pennsylvania and some of New Jersey."

Scheduling

With electric energy being generated by cogeneration plants, non-utilities, and utilities, and transmitted over a network of lines owned by generation companies, power pools, transmission companies, and various utilities, scheduling the delivery of energy becomes ever more complicated as market complexity and demand increase.

The original grids or power pools were built only to allow neighboring utilities to share resources. As demand outstripped the generation capacity of local utilities, other suppliers, local and distant, became connected to transmission lines in limited ways. On a daily and seasonal basis, intermediate and peak load generating plants are brought on-line as demand increases and taken off as it falls. An "event" at a nuclear power plant could take up to 10% of a region's generating capacity off line in an instant. While local peak load steam turbine generators can be spun up in minutes, occurrences such as this must be carefully anticipated. Generating plants owned by utilities and independent power producers negotiate long- and short-term contracts, with provisions for emergencies.

Substations

Both transmission and subtransmission lines terminate at substations (Fig. 18-12). If you think transmission lines with their conductors, insulators, towers, and UHVs are cool, there is even more interesting stuff in substations! While there are differences between the substations terminat-

Figure 18-12: Using existing substation avoids acquiring new rights-of-way

ing transmission and subtransmission lines, both will be described in this section.

Transformers are the heart of substations. Transformers convert the 115, 138, 230, 345, etc., kV transmission line voltage to the subtransmission voltage (typically 69 kV). In distribution substations, the 69 kV subtransmission voltage typically decreases to 13.8 kV. As always, there may be local or regional exceptions to these nominal voltages.

Invented in 1830 by American Joseph Henry, a transformer consists of primary and secondary windings and a laminated magnetic steel core. Transformers may increase or decrease current or voltage as required, but even though they are usually 97-99.5% efficient, the output power can never be greater than the input. Transformers may also be single or 3-phase. They may be air or oil cooled. Schematic diagrams of transformers normally show the primary and secondary windings on opposite sides of a squared ring representing the core. Actually, for reasons of compactness, efficiency and performance, the input voltage winding, called the primary, is normally wound around a bobbin or coil form and then the output voltage winding; the secondary, is wound over the primary.

Demand

While generating capacity can be brought on-line as needed to meet peak needs and substitute for nuclear units being refueled, etc., the transmission and distribution networks must be sized to carry the peak load.

A plethora of distribution voltages

If there are numerous transmission line voltages, there is a plethora of distribution voltages. For maximum efficiency, engineers design generators for the highest output voltage consistent with state of the art technology and customer needs. For many years, 13.8 kV was the standard. Now 20 or 22 kV is practical. However, when utilities were local entities, generators connected directly to the distribution network. If there was a generator built to operate on a given voltage, probably there is still a distribution voltage somewhere to match it. Thomas Edison's dynamos generated 110/220 dc. Those dc systems gave way to more efficient, higher-voltage ac distribution systems. With the longevity of industrial equipment in some localities, the once-mighty 4.2, 4.5 and 4.8 kV generators still serve as peaking units where a subtransmission substation carries the base and intermediate loads.

For the purposes of this book, the generic distribution voltage will be referred to as 13.8 kV. However, for any given utility, the distribution voltage may be as low as 4.2 kV. Subtransmission lines at 69 kV or even 138 kV transmission lines may also be considered to be part of a utility's distribution department. Note that a local utility's transmission, subtransmission, and distribution lines do not show up on "grid" maps.

"Grid? What grid?" Reprise

Just as the "grid" has serious limitations on the transmission side, not every customer can be served by redundant feeders for corporate, economic and practical reasons.

Let's return to Figures 18-2a, 2b, 2c. Fig. 18-2a pictures the end of line for WMEC (NE Utilities) distribution on Route 9 in Amherst, MA. Note that while the telephone lines continue to the pole shown in Fig. 18-2b, Massachusetts Electric's (NEES) 13.8 kV 3-phase lines on the crossbar feed an automotive repair business across the road and do not connect with

WMEC's distribution in Amherst. The 13.8 kV 3-phase distribution switch in Fig. 18-2c sits atop another pole on Route 9, just west of this location, and is normally left open (off). When heavy snow, a lightning strike, or a vehicle accident interrupts the single line into the neighborhood, there is no automated rerouting and no grid to assure continued power. The power stays off until utility crews repair the problem.

120/240 V distribution

Single transformers on utility poles reduce the nominal 13.8 kV distribution to 120/240 V single-phase for residential service. This scheme goes back to Edison's concerns for safety. In Europe and throughout much of the world, residential distribution uses 220 vac derived from 380 V 3-phase. While our current system arguably is safer, the European system is more cost-effective. Unfortunately, while many commercial and small industrial sites use 120/208 V single/3-phase systems, existing domestic clothes dryers, electric ranges, and water well pumps would not run efficiently on 208 V. A 277/480 V system would require replacing all existing electrical equipment except commercial fluorescent lighting. Therefore, no change is likely in the foreseeable future.

Part 5: Ecological and Environmental Considerations and Safety

Chapter 19
Cultural Aspects

In the year 1 AD, 200 million human beings walked the earth. More than 1,500 years passed before the world's population reached 500 million people. In 200 years, from 1650 to 1850, the population doubled to 1 billion. After only 80 more years, world population reached 2 billion. In 45 years the human population expanded to 4 billion and by 1997 was 5.8 billion souls.

For three quarters of history since Biblical times, population growth was nearly flat. In this century—some time before 1930—the curve turned up dramatically. Today, it is a nearly vertical line "approaching the asymptote," as a mathematician might say. (Fig 19-1)

Obviously, the earth cannot support an infinite population of energy- and food-hungry human beings. Already population growth is slowing in most of the more developed countries—0.6% in Canada, 1.0% in China, 0.1% in Denmark, 0.2% in France, 0.0% in Italy, 0.6% in the U.S.—and has turned negative in countries such as Belarus, Bulgaria, Croatia, Estonia, Germany, Hungary, Romania, and Russia. According to current projections, world population will peak and then begin to decrease after 2050. By then the total population may have doubled more than once.

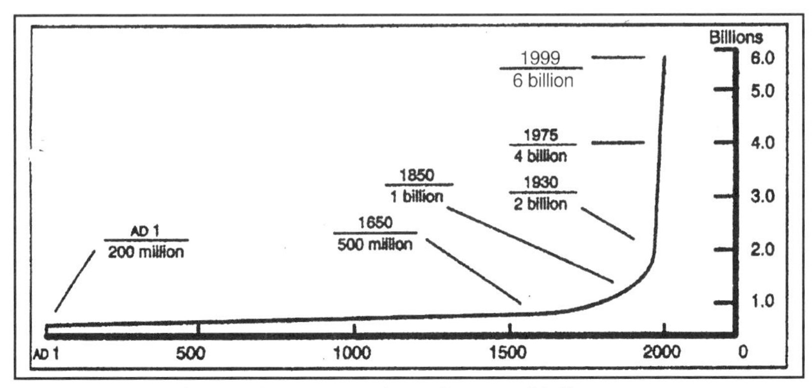

Figure 19-1: World Population Growth: "Billions and billions" Reprinted with permission from *The World Almanac and Book of Facts 1999*. Copyright © 1998 World Almanac Education Group. All rights reserved.

Looking at the spike in world population growth in this century, it is difficult to see how growth could simply flatten out again. History is full of famine, pestilence, plagues, and war that decimated populations—and it will take our best efforts to thwart the "Four Horsemen of the Apocalypse." However, the U.S., former Soviet Union, our European allies, and China managed to make it through the Cold War without turning the world into a radioactive cinder.

The odds are we will make it through this population crisis. We have to. Therefore, among other areas of concern, we must plan for the energy needs of a world of more than 10 billion people.

Security: Building a Sense of Us

Is security the armed ex-Navy Seal in the dark blue uniform behind the motorized gate who challenges anyone trying to enter a nuclear power plant? (Fig. 19-2) Obviously, someone needs to be on guard to initiate emergency protocols in the event of an accident, or criminal or terrorist action. In a critical industry, just having a uniformed presence reminds employees, vendors, and visitors that complying with procedures is not optional.

However, in an increasingly complex world where the potential list of alcohol, drugs, labor disputes, terrorist threats, and unspecified alienation problems continues to grow, the security guards at the front desk and gates must be thought of as the *last* line of defense, not the first. Threats to the secu-

Figure 19-2: First and last line of defense

rity of a utility or any organization come in many forms and on many levels.

Today's headlines make it easy to put terrorism at the top of the list of threats to security. Terrorism is a distinct and potentially real threat that must be planned for. However, dealing with more mundane problems on a continuing basis increases the likelihood of surviving a terrorist attack if it comes.

The human species forms complex social hierarchies and organizations. For healthy and pathological reasons, we divide ourselves rather quickly into "us" and "them." When things get bad, it becomes, us versus them: black vs. white, conservatives vs. liberals, civilian vs. military, criminals vs. society, Democrats vs. Republicans, free enterprise vs. government control, gay vs. straight, labor vs. management, men vs. women, government vs. religion, developers vs. environmentalists, poor vs. rich, union vs. exempt, and alcoholics and the drug dependent vs. everybody. "Honor diversity" may be dismissed as a catch phrase, or used by people of good will with differing needs and wants for building a larger sense of us. At least as many of our problems arise from our similarities as from our differences.

For instance, we all want what is best for our children. We may disagree vehemently on what that means. When we identify as members of a larger group, we can more easily and productively work together. Employees accept their differences with management—without abandoning them—if, on the whole, they feel everyone is working for a common goal. Members of the community concerned about the environment may accept the placement or design of cooling water intakes along a river or transmission structures if they feel every effort is being made to protect delicate habitats and aesthetics—and that the utility provides good jobs for friends, neighbors, and relatives. Some issues simply do not go away and must be dealt with on an ongoing basis.

During the televised commentary after William Bennett was installed as the nation's first "drug czar," a network consultant described alcohol as the biggest drug problem in the U.S., and cocaine as our biggest crime problem. Alcohol and drug problems persist and must be dealt with on a continuing basis. If substance abuse is treated as a disease, the blame and guilt can be dispensed with and the problem dealt with effectively. Like diabetes, alcohol and drug dependency is not the individual's fault; however, it is his or her responsibility to deal with it. Those impaired can not be allowed to endanger themselves, other employees or the public—or to steal assets to pay for drugs. Employee assistance programs (EAP) have a good track record of sorting out these issues. For a dollar or quarter in the basket to cover cost of coffee and room rental, 12-step programs often help people find recovery that insurance-provided therapy can not match. Interventions and professional therapy sometimes are also necessary and every effort needs to be made to negotiate this coverage if only to protect a company's investment in valuable employees.

Fears about the environment increase. "The Chicken Little World Tour" ("The sky is falling, the sky is falling!") began with the mercury-in-tuna fish scare in the mid-60s. The world tour of fear then visited every kind of shortage from energy to toilet paper. Then it was on to nuclear winter, global warming, el Niño and concern that the contrails of jet aircraft cause excessive cloud formation with unknown but bad results. Many "environmentalists" are well meaning and practical; a minority are anti-human and extreme in their demands to reduce consumerism and private ownership of internal combustion engine vehicles. What is the average person to believe? While global warming remains a potential threat, the nearly vertical rise in world population in this century can only increase the friction caused by billions more people rubbing elbows and competing for limited energy, food, potable water, and other resources.

"Think globally, act locally," exhorts a popular bumper sticker. What can a utility with stockholders and a bottom line to worry about do to remain financially viable and secure? "Think globally, but think," might be the appropriate answer. "Love thy neighbor," is an older answer.

No amount of concessions to radical environmentalists or boardroom despots will satisfy them. However, building "a sense of us" with employees, the surrounding community, and with regulators at the federal, local, and state level through consistent, responsible operations and sensitivity to legitimate concerns creates relationships in which individuals feel secure reporting potential problems before they become disasters.

With the Cold War over, terrorism continues almost as a bad habit. The original purpose of terrorism was to disrupt civilian life with destruction of commerce or infrastructure, such as electric power and water supplies, to force an already authoritarian government to increase security restrictions on its people. With Marxism repudiated by the peoples of most countries in the world and their governments it is difficult to view state-sponsored terrorism as having any purpose. What protection is there against such terrorism? How do we build a society secure enough that when there are concerns or problems, community members, employees, and managers alike feel free to voice them in an appropriate forum and a timely fashion?

Figure 19-3: Ben Jones with electric conversion Jetta.

As recent events have shown, disgruntled employees are not confined to the postal service. Treating people like the enemy may make them just that. Building that hard-to-define "sense of us" is the best way to deal with and defuse external and internal threats. It is a challenge that must be met.

Electric Cars: One Small Step?

Electric cars have been around as long as gasoline powered ones. Mrs. Henry Ford is said to have preferred her electric to the cars her husband manufactured. "Electric car offers guilt-free mileage," declares the headline in an issue of the *Weekend Gazette* (11-12 July 1998) "With a meter on the charger they know it costs only about six cents a mile to run." The converted Volkswagen Jetta delivers the promise of a pollution-free electric car. (Fig. 19-3) Or does it?

Let's compare the cost of electricity vs. gasoline to operate the vehicle. Recent VW Jettas are rated 23-24 miles per gallon (mpg) city and 30-31 mpg highway. Supposing the worst fuel efficiency number and a typical $1.059 per gallon for unleaded regular, it would cost 4.6 cents per mile to run a Jetta on gasoline. Supposing that other costs, such as insurance and maintenance, are the same (although electric motors tend to be more reli-

Figure 19-4: More batteries in the "boot"

Figure 19-5: Electric vehicle controls

able, with much lower maintenance than internal combustion engines), it's 30% more expensive to run the vehicle on electricity.

OK, that wasn't the question, was it? Does the electric car deliver on the promise of pollution-free transportation? No antifreeze to flush or oil to change—that more than compensates for the higher cost of electric energy, according to Dr. Emlen Jones, who uses the Jetta to commute to the HMO where he works as a pediatrician and to the hospital 10 miles away. (Come on, answer the question.) Except for trace amounts of ozone produced by the electric motor's brushes arcing on the commutator plates, essentially no pollution is produced by the vehicle. No greenhouse gases such as CO_2, NO_X or SO_2. Doesn't that settle it? Electric cars must be the answer—right?

Not so fast. Electric heat is 100% efficient. So why is electric heat the most expensive way to heat a building? Electric generators (and motors) are very efficient, say 70%. (The overall efficiency of the most modern natural gas-fired combined-cycle turbine generating plants is about 60%.) Transformers are up to 97% efficient. The first transformer in the switchyard of the generating plant steps up the generator output to transmission line voltage of about 138,000 V. At a transmission substation, a second transformer reduces the voltage to a branch circuit to 69,000 V. At a third transformer in a distribution substation, the voltage drops to 13,800 V. Finally the pole transformer across the street from your house, converts the 13.8 kV to the 120/240 V you use for lights, appliances—and the charger for your electric car. Let's call transmission lines 99% efficient per mile and you are 20 miles from the nearest generating plant.

Next, consider the efficiency of the electric generating, transmission and distribution system in Table 19-1.

Description	Efficiency	Number of elements/circuit
Combined Cycle Turbine Generator	.60	1
Transformer	.97	4
Transmission Line	.99/mile	20
Battery	.60 (Charge/Discharge)	1 Charge, 1 Discharge
Motor	.70	1

Table 19-1: Generation, Transmission and Distribution Efficiency

An automotive gasoline engine is about 15% efficient. Given that the scrubbers on a coal-fired power plant are more efficient than the emission controls on an automobile, even the lower-efficiency electric vehicle is somewhat less polluting than the gas-powered car. If the electric car's batteries are recharged at night when the base load is being carried by hydro and nuclear plants, it could be said to be non-polluting in terms of CO_2, NO_x, SO_2 and water vapor emissions (Fig. 19-4). However, the heat load from the nuclear plant is still greater than that of a conventional vehicle.

While the converted Jetta rides comfortably and has good acceleration on level road, it lags going up hills (to the annoyance of drivers following). With its gasoline engine removed and batteries and electric motor installed, it is about 500 pounds heavier than when it left the VW factory. Within 100 pounds of its maximum load rating, it is like car pooling with three other people even when the driver is alone (Fig. 19-5).

From an environmental and safety viewpoint—and although the lead-acid batteries are encased in sturdy battery boxes—in a severe collision sulfuric acid and lead could be released, burning the vehicle occupants and contaminating the surroundings.

Except where vehicle emissions are both an acute and chronic problem, the benefits of electric vehicles over conventional ones are marginal, therefore public policy decisions in favor of electric vehicles are not warranted.

Chapter 20
Safety First, Last, and Always

Why do we need safety when we have lawyers?

— Anon.

Your color television set or computer monitor develops about 30,000 V at less than twenty-thousandths of an A to accelerate electrons at the screen and produce the images you see. If you come in contact with this voltage, enough current would be forced into your body to make your muscles contract violently and throw you across a room. However, unless you are as unlucky as you were careless, you would probably live to tell the tale. Depending on how you landed, you might even find the experience exhilarating.

The final amplifier tubes (yes, tubes—big tubes!) of a modern UHF television transmitter operate with approximately 30,000 V between the cathode (where electrons are launched into the vacuum) and water-cooled collector (which returns them to the power supply), averaging about two As current. Control interlocks, safety earth grounding switches, and some extremely fast electronic "crowbar" circuitry make it unlikely you could cause a short circuit—these tubes are expensive, after all, and time off the air is even more costly to a TV station. However, if you are persistent, the

Figure 20-1: A bright idea—safety first!

Figure 20-2: A commitment to safety

30,000 V power supply can deliver about 200 As in a pinch. You're dead. Unless there is someone trained in cardiopulmonary resuscitation (CPR) on the scene and advanced life support only a few minutes away—and your luck changes drastically—you are going to stay dead.

Figure 20-3: Sound-absorbing walls protect neighbors from and repair noise—the plant itself is quiet.

A 50 MW generator operating at 30,000 V (20,000 V is more typical but don't stop us, were on a roll!) delivers almost 1,700 As—as much as an average lightning bolt. However, a lightning strike lasts a fraction of a second, while the generator just keeps on turning. Now you're not just dead—you're carbon.

In spite of the fact this is the last chapter in the book—Safety First.

Today, government environmental health and safety regulations override organizational, managerial, or individual intentions with evermore-complex rules, even as bureaucrats pay lip service to paperwork reduction and reinventing government. In the end, an effective environmental health and safety program comes back to the proper motivation: doing the right thing for the right reasons on the individual plant level.

Reading this book will not qualify you to deal with a downed wire—treat any downed lines as deadly. Keep yourself, co-workers, family, friends, and any strangers who will listen well away from them. Because metal guardrails are imperfect electrical grounds, downed wires in contact with them may make them lethal. Just because ambulance emergency medical technicians (EMTs), firefighters, and police are on the scene does not mean it is safe to approach downed wires. In their excitement to reach

Figure 20-4: Security card access

accident victims, burning vehicles, or to control the scene, they may neglect their training and put themselves in grave danger. The experts in this situation are the hardhatted employees arriving in the truck with the flashing yellow lights. They know the safest place for you to be is home

Figure 20-1: Conventional but top of the line smoke and thermal detection system

watching live coverage of the incident on TV.

If you really want to help, start by taking a CPR course. It teaches you not just several ways to save lives, but strategies to prevent death, disease, and injury. You will also learn how to activate your area's emergency medical system (EMS) so you're not a hero alone in the wilderness and risk becoming a casualty yourself. CPR is the gateway to first responder or standard and advanced first aid courses or even EMT training.

Electricity can be deadly, but its behavior is entirely predictable—unlike the effects of donuts, french fries, smoking cigarettes, or lack of regular exercise!

Participate whole-heartedly in environmental, safety, and security training programs when offered. If you gotta attend, you might as well learn something useful. If you are in a leadership or management position, take the lead! Be part of the committee that develops and prototypes the program, or at least be in the first regular class. You did not get where you are by neglecting the nitty-gritty details of your field. Environmental, safety, and security programs are everyone's field. If you are a worker bee, show those managers you are not just a drone! The whole organization benefits if you keep them on their toes.

To paraphrase an old but true slogan-a healthy environment, safety, and security are no accident.

Appendix

Net Generation from U.S. Electric Utilities by Energy Source, Census Division, and State

1997 and 1998 (Million kWh)

Census Division State	Coal		Petroleum[1]		Gas	
	1997	1998	1997	1998	1997	1998
New England	19,124	13,164	22,494	21,759	10,340	4,859
Connecticut	2,558	1,483	8,431	8,608	1,546	977
Maine	—	—	1,443	1,729	—	—
Mass.	12,489	8,169	11,586	10,020	5,213	1,819
New Hampshire	4,077	3,513	1,008	1,353	35	10
Rhode Island	—	—	17	9	3,546	2,053
Vermont	—	—	10	41	*	1
Middle Atlantic	134,019	135,607	10,834	19,106	24,094	23,339
New Jersey	6,822	5,586	384	485	2,777	2,854
New York	21,752	23,503	8,142	14,524	20,706	19,913
Pennsylvania	105,446	106,517	2,307	4,097	611	572

East North Central	416,285	418,627	2,147	3,216	5,996	9,117
Illinois	76,092	70,306	495	838	3,442	4,483
Indiana	108,912	110,696	607	822	386	775
Michigan	65,552	69,143	602	1,005	838	2,152
Ohio	124,910	128,696	273	351	228	519
Wisconsin	40,820	39,786	170	200	1,101	1,188
West North Central	189,797	201,886	1,204	1,307	3,749	5,832
Iowa	28,739	31,884	82	110	277	412
Kansas	27,236	28,024	110	122	2,068	2,924
Minnesota	27,081	29,884	764	650	512	652
Missouri	59,903	62,489	125	310	570	1,232
Nebraska	17,209	18,336	31	42	206	400
North Dakota	26,314	28,176	86	47	*	*
South Dakota	3,314	3,094	7	27	117	211
South Atlantic	382,150	390,087	29,754	49,880	38,136	39,397
Delaware	3,926	3,812	833	1,234	1,820	1,272
District of Columbia	—	—	71	244	—	—
Florida	66,035	65,470	25,742	40,953	32,998	31,711
Georgia	66,180	69,871	201	671	568	1,769
Maryland	27,394	29,077	1,479	3,312	879	1,054
North Carolina	70,181	69,001	212	286	377	936
South Carolina	31,043	32,378	188	331	181	415
Virginia	29,676	31,471	858	2,655	1,292	2,199
West Virginia	87,715	89,008	171	194	21	42
East South Central	230,861	220,738	3,070	6,504	6,495	9,131
Alabama	71,586	71,457	119	260	885	2,449
Kentucky	87,875	82,412	126	127	177	496
Mississippi	12,501	11,748	2,633	5,418	5,281	5,635
Tennessee	58,899	55,120	193	699	152	551
West South	212,447	207,556	913	888	142,924	169,222
Arkansas	22,761	23,140	67	144	2,243	3,704
Louisiana	20,953	20,762	646	600	26,010	28,318
Oklahoma	33,037	31,027	13	8	12,507	17,000

Texas	135,696	132,627	188	137	102,164	120,201
Mountain	194,420	207,005	233	260	11,058	14,788
Arizona	34,219	36,225	61	61	2,065	3,472
Colorado	32,002	33,079	15	37	424	964
Idaho	—	—	*	*	—	—
Montana	14,410	16,508	17	14	32	41
Nevada	15,251	17,161	31	50	5,021	6,190
New Mexico	27,079	27,537	21	23	3,210	3,631
Utah	32,144	33,207	29	31	297	463
Wyoming	39,315	43,287	59	43	10	27
Pacific Contiguous	8,467	12,639	169	193	37,803	30,988
California	—	—	142	121	36,301	26,385
Oregon	1,501	3,348	11	33	1,273	3,467
Washington	6,966	9,290	16	39	229	1,135
Pacific Noncontiguous	237	171	6,935	7,044	3,031	2,549
Alaska	237	171	741	757	3,031	2,549
Hawaii	—	—	6,194	6,287	—	—
U.S. Total	1,787,806	1,807,480	77,753	110,158	283,625	309,222

	Nuclear		Hydroelectric[2]		Renewable[3]	
	1997	1998	1997	1998	1997	1998
New England	16,432	20,686	4,508	4,359	601	573
Connecticut	-125	3,243	367	385	451	427
Maine	0	0	1,780	1,820	—	—
Massachusetts	4,310	5,698	300	331	—	—
New Hampshire	7,979	8,387	1,165	975	—	—
Rhode Island	—	—	0	0	—	—
Vermont	4,267	3,358	896	848	150	145
Middle Atlantic	111,132	119,595	28,930	28,004	18	5
New Jersey	13,908	27,132	-130	-146	—	—
New York	29,570	31,314	27,912	26,582	18	5
Pennsylvania	67,655	61,149	1,148	1,568	—	—
East North Central	92,229	93,963	3,926	2,806	395	441

Illinois	51,069	55,596	17	51	24	0
Indiana	—	—	562	479	—	—
Michigan	21,914	12,494	658	352	—	—
Ohio	15,331	16,476	507	406	—	—
Wisconsin	3,916	9,397	2,182	1,518	372	441
West North Central	41,622	42,598	16,975	13,593	494	549
Iowa	4,149	3,768	795	893	22	19
Kansas	8,430	10,411	—	—	—	—
Minnesota	10,819	11,644	697	695	429	451
Missouri	8,955	8,517	1,478	2,269	42	78
Nebraska	9,269	8,259	1,672	1,683	1	1
North Dakota	—	—	3,320	2,296	—	—
South Dakota	—	—	9,012	5,758	—	—
South Atlantic	171,048	190,598	12,895	14,205	0	0
Delaware	—	—	—	—	—	—
District of Columbia	—	—	—	—	—	—
Florida	22,968	31,115	241	199	—	—
Georgia	30,414	31,380	4,418	5,026	—	—
Maryland	13,213	13,331	1,588	1,740	—	—
North Carolina	32,453	38,778	4,148	4,111	—	—
South Carolina	44,916	48,759	2,047	2,513	—	—
Virginia	27,084	27,234	76	256	0	0
West Virginia	—	—	377	361	—	—
East South Central	65,033	66,241	24,302	23,066	—	—
Alabama	29,573	28,663	11,521	10,565	—	—
Kentucky	—	—	3,380	3,116	—	—
Mississippi	10,813	9,191	—	—	—	—
Tennessee	24,648	28,388	9,401	9,385	—	—
West South Central	65,077	68,210	8,120	7,953	*	*
Arkansas	14,208	13,097	3,511	3,114	—	—
Louisiana	13,511	16,428	—	—	—	—
Oklahoma	—	—	2,824	3,420	—	—

Texas	37,358	38,685	1,785	1,419	*	*
Mountain	29,314	30,301	46,735	41,692	169	160
Arizona	29,314	30,301	12,401	11,239	—	—
Colorado	—	—	1,935	1,392	0	0
Idaho	—	—	13,512	11,978	—	—
Montana	—	—	13,348	11,054	—	—
Nevada	—	—	2,567	3,151	—	—
New Mexico	—	—	259	236	—	—
Utah	—	—	1,331	1,299	169	160
Wyoming	—	—	1,381	1,342	—	—
Pacific Contiguous	36,756	41,510	189,725	167,598	5,785	5,478
California	30,512	34,594	39,797	48,684	5,431	5,141
Oregon	—	—	46,283	39,504	0	0
Washington	6,244	6,916	103,645	79,410	353	337
Pacific Noncontiguous	—	—	1,118	1,127	—	*
Alaska	—	—	1,099	1,113	—	—
Hawaii	—	—	19	14	—	*
U.S. Total	628,644	673,702	337,234	304,403	7,462	7,206

Notes

[1] Includes petroleum coke.

[2] Station losses include energy used for pumped storage. Energy used in 1998 for pumping was 28,872 million kWh and in 1997 was 28,342 million kWh.

[3] Includes geothermal, biomass, wind, solar, thermal and PV (excludes hydroelectric).

= Value less than 0.5 million kWh.

Notes:

- Data are final
- Negative generation denotes that electric power consumed for plant use exceeds gross generation
- Totals may not equal sum of components because of independent rounding

Source: Energy Information Administration, Form EIA-759, "Monthly Power Plant Report."

Glossary

A

acid rain. Rain with a pH below 5.6. Rain normally has a pH of around 5.6, which is slightly acidic. Rain becomes more acidic when nitric and sulfuric acids from nitrogen oxides and sulfur oxides are released into the atmosphere. One of the most common ways these oxides are released is through the burning of fossil fuels.

actuator. A controlled motor that converts electricity to action, or any device that converts voltage or current into a mechanical output.

adiabatic. Any change in which there is no gain or loss of heat.

affiliated power producer. Electric power generated by a facility that is affiliated with another facility, which also supplies such services or equipment.

air monitoring. Intermittent or continuous testing of emission air for pollution levels.

air pollution. Contaminants in the atmosphere, which have toxic characteristics and which are believed to be harmful to the health of animal or plant life.

air quality. Air quality is determined by the amount of pollutants and contaminants present.

Alkaline. A compound dissolved in water with a pH greater than seven; a base.

allowable emissions. The emissions rate of a stationary source calculated using the maximum rated capacity of the source and the most stringent applicable governmental standards.

alternating current. (ac) A periodic current, the average value of which over a period is zero. Unless distinctly specified otherwise, the term refers to a current that reverses its direction at regularly recurring intervals of time and that has alternately positive and negative values. Almost all electric utilities generate AC electricity because it can easily be transformed to higher or lower voltages.

ambient conditions. The outside weather conditions, including temperature, humidity, and barometric pressure. Ambient conditions can affect the available capacity of a power plant.

American wire gauge. Abbreviated AWG. The standard wire size measuring system in the U.S. Higher numbers indicate smaller wires.

ammeter. A device used to measure an electric current's magnitude, by amperes, milliamperes, microamperes, or kiloamperes.

ampere. (A) A unit of measurement of electric current produced in a circuit by 1 Volt acting through a resistance of 1 Ohm.

anthracite. A hard, black lustrous coal, often referred to as hard coal. Anthracite contains a high percentage of fixed carbon and a low percentage of volatile matter.

APPA. American Public Power Association. The trade association of publicly held power entities.

arc. A discharge of electricity through a gas or air

ash. Impurities consisting of silica, iron, aluminum, and other noncombustible matter that are contained in coal. Ash increases the weight of coal, adds to the cost of handling, and can affect its burning characteristics.

atom. The smallest particle of an element having the chemical properties of that element. It is the fundamental building block of elements.

attainment area. A geographic area under the Clean Air Act, which is in compliance with the Act's national Ambient Air Quality Standards. This designation is made on a pollution-specific basis.

availability. The unit of measure for the actual time a generating unit is capable of providing service, if needed.

avoided cost. A utility company's production or transmission cost avoided by conservation or purchasing from another source rather than by building a new generation facility.

B

background radiation. Also called natural radiation; it is the radiation that is always present in the environment from natural sources such as cosmic rays, and radioactive elements in the ground, building materials, and the human body.

back-up power. Power supplied to a customer when its normal supply is interrupted.

barrel. A volumetric unit of measure for crude oil and petroleum products equivalent to 42 U.S. gallons.

base load. The minimum amount of electric power or natural gas delivered or required over a given period of time at a steady rate.

base load plant. A power plant, usually housing high-efficiency steam-electric units, which is normally operated to take all or part of the minimum load of a system and which consequently produces electricity at an essentially constant rate and runs continuously. These units are operated to maximize system mechanical and thermal efficiency and to minimize system operating costs.

Biomass. Any organic matter that can be used as a fuel to generate energy. Wood and wood waste are common examples of biomass fuel, but biomass also includes such items as agricultural wastes, lawn and yard waste, and animal waste. All of these can be converted to energy-producing fuels.

bituminous coal. The most common coal. It is dense and black, often with well-defined bands of materials. Bituminous coal is used for generating electricity, making coke, and for space heating.

blade. In wind generation, an aerodynamic surface which extracts energy from the wind.

boiler. A device for generating steam for power, processing, or heating purposes or for producing hot water for heating purposes or hot water supply. Heat from an external combustion source is transmitted to a fluid contained within the tubes in the boiler shell. This fluid is delivered to an end-user at a desired pressure, temperature, and quality.

boiling water reactor. Abbreviated BWR. A nuclear power reactor cooled and moderated by light water (regular water), which is allowed to boil in the core, generating steam for the turbine.

Boron. A chemical element that absorbs neutrons. Boron can be used to control or stop a chain reaction in a nuclear reactor.

bottom ash. Also called cinders. It is a granular by-product of burning coal to generate electricity.

BPA. Bonneville Power Administration. A power marketing and electric transmission agency of the U.S. government with headquarters in Portland, Oregon.

breaker. See *circuit breaker*

breeder reactor. A reactor that produces more fissionable material than it consumes.

Btu. British thermal unit. A standard unit for measuring the quantity of heat energy equal to the quantity of heat required to raise the temperature of a pound of water by 1°F.

bulk power. The generation and high-voltage transmission of electricity.

bundling. Combining the costs of generation, transmission, and distribution and other services into a single rate charged to the retail customer.

bus. A conductor or solid bar aluminum or copper used to connect a circuit or circuits to a common interface. For example, the bus bars or conductors in a substation used to connect the low-voltage windings of transformers to the outgoing distribution circuits.

busbar cost. The traditional measure for the total cost of generating electricity. Busbar cost refers to the cost of generated electricity before it enters the utility's system.

bushing. An insulating structure including a through conductor with provision for mounting on electrical equipment for the purpose of insulating the conductor from its mounting and conducting current through the mounting.

C

cable. A conductor with insulation or a stranded conductor with or without insulation and other coverings or a combination of conductors insulated from one another.

capability. The maximum load that a generating unit, generating station, or other electrical apparatus can carry under specified conditions for a given period of time without exceeding approved limits of temperature and stress.

capacitance. The property of a system of conductors and dielectrics that permits the storage of electricity when potential differences exist between the conductors.

capacitor. A device that stores a dc charge and passes an ac current. Used to smooth dc power supplies and to stabilize ac line voltages.

capacitor bank. An assembly of capacitors and all necessary accessories, such as switching equipment, protective equipment, controls, and other devices needed for a complete operating installation.

capacity. The amount of electric power delivered or required for which a generator turbine, transformer, transmission circuit, station, or system is rated by the manufacturer.

capacity charge. An element in a two-part pricing method used in capacity transactions. Energy charge is the other element. The capacity charge, also called a demand charge, is assessed on the amount of capacity being purchased.

chain reaction. The process by which atomic fission becomes self sustaining. A neutron fired at the nucleus of an atom causes fission, which releases other neutrons, which then hit other nuclei.

carbon dioxide. A colorless, odorless, nonpoisonous gas which occurs in ambient air. It is produced by fossil fuel combustion or the decay of materials.

carbon monoxide. A colorless, odorless, tasteless, but poisonous gas produced mainly from the incomplete combustion of fossil fuels.

charge. A property in some elementary particles which causes them to exert forces on each other. The force is thought to result from the exchange of photons between the charged particles. Charge is the result of an excess or deficiency in electrons in respect to protons present.

circuit. A circuit consists of an electrical power source, supply and return conductors, and a load or loads through which an electric current flows from the source through the conductor to the load, through the load, and back to the source.

circuit breaker. A device designed to open and close a circuit. It is designed to open automatically on a predetermined overload of current, without injury to itself, when properly applied within its rating.

circuit recloser. A line protection device that interrupts momentary line faults in a distribution system. A circuit recloser will automatically close after a short time and will immediately reopen the circuit if there are still problems.

coal. A black or brownish-black solid combustible substance formed by the partial decomposition of vegetable matter without access to air. The rank of coal which includes anthracite, bituminous coal, subbituminous coal, and lignite, is based on fixed carbon, volatile matter, and heating value. Coal rank indicates the progress alteration from lignite to anthracite. Lignite contains approximately 9 to 17 million Btus per ton. The contents of subbituminous and bituminous coal range from 16 to 24 million Btus per ton, and from 19 to 30 million Btus per ton, respectively. Anthracite contains approximately 22 to 28 million Btus per ton.

cogeneration. The simultaneous production of power and thermal energy, such as burning natural gas to produce electricity and using the heat produced to create steam for industrial use.

combined-cycle. An electric generating technology in which additional electricity is produced from otherwise lost waste heat exiting from the gas turbines. The exiting heat is routed to a conventional boiler or to a heat recovery steam generator for utilization by a steam turbine in the production of electricity. The process increases the efficiency of the electricity generating unit.

combustion. The rapid chemical combination of a substance with oxygen, usually accompanied by the liberation of heat and light.

combustion air. Air needed to ensure complete fuel combustion.

combustion chamber. The area where fuel is burned.

condenser. Condensers are equipment in generating facilities that capture steam and turn it back into water for reuse in the feedwater system of the plant.

conductor. A material, usually in the form of a wire, cable, or bus bar, suitable for carrying an electric current.

conduit. A structure designed to hold electric conductors. It could be a metal pipe or other material.

connector. A coupling device used to connect conductors of one circuit to those of another.

consumption. The amounts of fuel used for gross generation, providing stanby service, start-up, and/or flame stabilization. May also be used to refer to customer use.

containment. The structures, including the reactor building, designed to prevent the escape of radiation from a nuclear reactor to the outside environment.

control rods. Rods made of neutron absorbing material such as cadmium that are used to regulate or stop nuclear fission in a reactor.

convergence. The coming together and merging of previously distinct industries. This phenomenon is currently under way for the electricity and fuels industries, particularly electricity and natural gas.

cooling tower. The portion of a power facility's water circulating system which extracts the heat from water coming out of the plant's condenser, cooling it down and transferring the heat into the air while the water returns through the system to become boiler make-up water.

core. The central portion of a nuclear reactor containing the fuel rods, moderator, and control rods where nuclear fission takes place in the core.

critical. Term used to describe a nuclear reactor that is sustaining a chain reaction.

critical mass. For a given geometry (shape: cube, cylinder, rectangle, sloid, spherical, etc.) the minimum amount of fissionable material needed to start and continue a chain reaction in nuclear reactor.

cubic foot. The most common unit of measurement of gas volume, it is the amount of gas required to fill a volume of one cubic foot under stated conditions of temperature, pressure, and water vapor.

current. The flow of electrons in an electrical conductor. The rate of movement of the electricity, measured in amperes.

D

decommissioning. The process of closing down a nuclear reactor after its useful life is ended.

decontamination. The removal of radioactive material.

demand. The rate at which electric energy is delivered to or by a system, part of a system, or piece of equipment at a given instant or averaged over a designated period of time.

demand-side management. Abbreviated DSM. The term for all activities or programs undertaken by an electric system or its customers to influence the amount and timing of electricity use. Included in DSM are the planning, implementation, and monitoring of utility activities that are designed to influence consumers use of electricity in ways that will produce desired changes in a utility's load shape. These programs are dwindling, and expected to experience a great decline under deregulation.

Department of Energy. Abbreviated DOE. Established in 1977, the DOE manages programs of research, development and commercialization for various energy technologies, and associated environmental, regulatory, and defense programs. DOE promulgates energy policies and acts as a principal adviser to the President on energy matters.

deregulation. Relaxing or eliminating laws and regulations controlling an industry or industries.

deuterium. (D or H^2) A "heavy" hydrogen nucleus consisting of a proton and a neutron. Used in heavy water (D_2O) moderated reactors such as the CANDU.

direct current. (dc) An electric current that flows in one direction with a magnitude that does not vary or that varies only slightly.

distillate fuel oil. A general classification for one of the petroleum fractions produced in conventional distillation operations. It is used primarily for space heating, diesel engine fuel, and electric power generation.

distribution company. An electric distribution company that provides only distribution services that are unbundled. Abbreviated DISCO.

distribution system. A term used to denote that part of an electric power system that distributes the electricity to consumers from a bulk-power location such as a substation. It includes all lines and components of the system beyond the substation fence.

draft. The movement of air into and through a combustion chamber, breeching, stack, and chimney. It can be natural, allowing hot air to rise, or artificial, produced by equipment such as fans.

E

Edison Electric Institute. Abbreviated EEI. The association of the investor-owned electric utilities in the U.S. and industry affiliates worldwide. Its U.S. members serve almost all of the customers served by the investor-owned segment of the electric utility industry. They generate almost 80% of all electricity generated by utilities and service more than 75% of all customers in the nation. EEI's basic objective is the "advancement in the public service of the art of producing, transmitting, and distributing electricity and the promotion of scientific research in such field". EEI compiles data and statistics relating to the industry and makes them available to member companies, the public, and government representatives.

electric capacity. The ability of a power plant to produce a given output of electric energy at an instant in time. Capacity is measured in kilowatts or megawatts.

electric current. A flow of electrons in an electrical conductor. The strength or rate of movement of the electricity is measured in amperes.

electric plant. A facility containing prime movers, electric generators, and auxiliary equipment for converting other types of energy into electric energy.

electric utility. A company that controls the distribution of electricity in a specific state, area, or region. Utilities often own and operate electricity generation, transmission, and distribution facilities.

electricity. The flow of electrons in a conducting material. The flow is called a current.

electromotive force. (EMF) The electrical pressure (voltage) exerted on a circut by an alternator, battery, dynamo, or generator. Measured in Volts (V).

emissions. Any waste products leaving a power plant. This term generally applies to air pollution, but it can also apply to soil or water waste issues. There are many substances that can be emitted from power plants, and most of them are regulated and monitored.

energy. Power is the capability of doing work. Energy is power supplied over time, expressed in kilowatt-hours. Energy can take on different forms, some of which are easily convertible and can be changed to another form useful for work. Most of the world's convertible energy comes from fossil fuels that are burned to produce heat that is then used as a transfer medium to medium to mechanical or other means in order to accomplish tasks. Electrical energy is usually measured in kilowatt-hours, while heat energy is generally measured in British thermal units.

energy charge. The portion of the charge for electric services that is based on the electric energy either consumed or billed.

energy marketer. An entity, regulated by the federal Energy Regulatory Commission, which arranges bulk power transactions for end users. The main goal of energy marketers is determining the best overall fuel choice for customers, whether it be natural gas, electricity, oil, etc., and then delivering that fuel to the customer. They deal in the open market, taking full title to the energy until they resell it to an end user.

Energy Policy Act of 1992. Legislation which authorized FERC to introduce competition at the wholesale level through new open access requirements for transmission and authorizing exempt wholesale generators.

Environmental Protection Agency. Abbreviated EPA. This agency administers federal environmental policies, enforces environmental laws and regulations, performs research, and provides information on environmental subjects.

exempt wholesale generator. A company that generates power solely for wholesale use and does not sell it to the public. They are exempt from PUHCA.

externalities. Factors affecting human welfare not included in the monetary cost of a product, such as air pollution caused by power generation.

extra high voltage. Abbreviated EHV. A term applied to voltage levels of electric power system transmission lines that are higher than 230,000 volts (230 kV).

F

Fahrenheit. Abbreviated F. The temperature scale commonly used in the U.S., with the freezing point of water at 32 degrees and the boiling point at 212 degrees at sea level.

fault. A partial or total local failure in the insulation or continuity of a conductor in a wire or cable.

federal electric utilities. A classification of utilities that applies to those that are agencies of the federal government involved in the generation and/or transmission of electricity. Most of the electricity generated by federal electric utilities is sold at wholesale prices to local government-owned and cooperatively owned utilities, and to investor-owned utilities. These government agencies are the Army Corps of Engineers and the Bureau of Reclamation, which generate electricity at federally owned hydroelectric projects. The Tennessee Valley Authority produces and transmits electricity in the Tennessee Valley region.

Federal Energy Regulatory Commission. Abbreviated FERC. Federal agency responsible for regulating interstate wholesale electricity markets and the interstate transmission of electricity.

Federal Power Act. This legislation was enacted in 1920 and amended in 1935. It has three parts. The first part incorporated the Federal Water Power Act, licensing nonfederal hydropower projects. The second and third parts were added with the passage of the Public Utility Act. The parts extended the Act's jurisdiction to include regulating the interstate transmission of electrical energy and rates for its sale as wholesale in interstate commerce. The Federal Energy Regulatory Commission is now charged with the administration of this law.

Federal Power Commission. The predecessor of the Federal Energy Regulatory Commission.

Federal Energy Regulatory Commission. Abbreviated FERC. The chief energy regulatory body of the U.S. government.

feeder cable. A cable which extends from a central site along a primary route or from a primary route to a secondary route, thus providing connections to one or more distribution cables.

feedwater. The water used in the boiler system of generating plants. It is treated to make it as pure as economically feasible to keep the boiler clean and operating properly.

firm service. Sales and/or transportation service provided without interruption throughout the year. Firm services are generally provided under filed rate tariffs.

fission. The splitting of atoms, resulting in the release of large amounts of energy. Fission can occur naturally, or when an atom's nucleus is bombarded by neutrons, as in a nuclear power plant.

fixed cost. An expense that does not change in response to varied production levels.

flue gas desulfurization unit. Also called a scrubber. Equipment used to remove sulfur oxides from the combustion gases of a boiler plant before discharge to the atmosphere. Chemicals are used to pull oxides from the gases.

flue gas particulate collectors. Equipment used to remove fly ash from the combustion gases of a boiler plant before discharge to the atmosphere. Particulate collectors include electrostatic precipitators, mechanical collectors, fabric filters or bag houses, and wet scrubbers.

fly ash. Very small diameter particle matter from coal ash which can "fly" on the gases emitting from the stacks of a power plant. Fly ash is generally removed from the flue gas by using flue gas particle collectors, such as fabric filters and electrostatic precipitators.

fossil fuel. Any naturally occurring organic fuel, including petroleum, coal, and natural gas.

fossil fuel plant. An electricity producing generating plant that uses coal, petroleum, and/or natural gas as its energy source.

fuel cell. A device that produces electrical energy directly from the controlled electrochemical oxidation of the fuel. It does not contain an intermediate heat cycle, as do most other electrical generation techniques.

fuel expenses. Costs include the fuel used in the production of steam and/or electricity at an electric power plant. Other associated costs include unloading the shipped fuel and all handling of the fuel up to the point where it enters the power plant. Fuel expenses are generally the largest expense category for electric power generating facilities.

fuse. A device that protects a circuit by fusing open its current-responsive element when an overcurrent or short-circuit current passes through it.

fusion. A nuclear reaction requiring exceedingly high temperatures and involving two light nuclei which fuse together forming one heavier nuclear, releasing large amounts of energy. The energy from the sun and most of the energy of hydrogen bombs come from fusion.

G

gamma ray. Ionizing, electromagnetic radiation of higher energy and shorter wavelength than x-rays.

gas turbine. Consists of an axial-flow air compressor and one or more combustion chambers where liquid or gaseous fuel is burned The hot gases that are produced are passed to the turbine and where the gases expand to drive the generator and are then used to run the compressor.

generating unit. Any combination of generators, reactors, boilers, combustion turbines, or other prime movers operated together or physically connected to produce electric power.

generation. The process of producing electric energy by transforming other forms of energy. It also refers to the amount of electric energy produced, generally expressed in kilowatt-hours or megawatt-hours.

generator. A machine that converts mechanical energy into electrical energy.

generator nameplate capacity. The full-load continuous rating of a generator, prime mover, or other electric power production equipment under specific conditions as designated by the manufacturer. Installed generator nameplate rating is usually indicated on a plate physically attached to the generator.

geothermal plant. An electric power plant in which the prime mover is a steam turbine. The turbine is driven either by steam produced from hot water or by natural steam that derives its energy from heat found in rocks or fluids at various depths beneath the surface of the earth. The energy is extracted by drilling and/or pumping.

gigawatt. Abbreviated GW. A unit of electric power equal to one billion watts or one thousand megawatts.

gigawatt-hour. Abbreviated GWh. One billion watt-hours.

Greenfield plant. Refers to a new electric power generating facility built from the ground up on a site that has not been used for industrial uses previously. Essentially a plant that starts with a green field. Plants that are built on sites that have already been used for another power plant or other industrial use are called brownfield plants.

greenhouse effect. The allegedly increasing mean global surface temperature of the earth, believed to be caused by gases in the atmosphere, including carbon dioxide, methane, nitrous oxide, ozone, and chlorofluorocarbons. The greenhouse gases allow solar radiation to penetrate but absorb the infrared radiation returning to space.

greenhouse gases. Those gases, such as carbon dioxide, nitrous oxide, and methane, that are transparent to solar radiation but opaque to longwave radiation. Their action in the atmosphere is similar to that of glass in a greenhouse.

gross generation. The total amount of electric energy produced by the generating units at a generating station or stations, measured at the generator terminals.

ground. A conducting connection, whether intentional or accidental, by which an electric circuit or equipment is connected to the earth, to some conducting body of relatively large extent such as structural steel in buildings or a vehicle chassis that serves in place of the earth.

H

half-life. The time required for half the atoms of a radioactive substance to fission (split) into lighter elements.

heat rate. A power plant term for the efficiency of the power plant. Heat rate measures how much of the fuel that is burned actually turns into electricity. Heat rate is generally represented as a mixture of British and metric units, Btu/kWh.

heavy oil. Fuel oils remaining after the lighter oils have been distilled off during the refining process. Except for start-up and flame stabilization, virtually all petroleum in steam plants is heavy oil.

Hertz. Abbreviated Hz. The international standard unit of frequency, defined as the frequency of a periodic phenomenon with a period of one second. Electricity is generally 60 Hz in North America and 50Hz in most countries other than Canada, Japan, and Mexico.

high-voltage system. An electric power system having a maximum root-mean-square ac voltage above 72.5 kilovolts.

hydrocarbon. An organic chemical compound of hydrogen and carbon in either gaseous, liquid, or solid phase. The molecular structure of hydrocarbon compounds varies from the simple, such as methane, to the very heavy and very complex.

hydroelectric pumped storage. Refers to a process for storing electric energy in the form of water held in an upper reservoir. Water is pumped from a lower reservoir into an upper reservoir during off-peak periods. It is then released through hydroelectric turbines into the lower reservoir for generation during peak-demand periods for management of electrical system loads, or to help provide needed power during system emergencies.

I

independent power producer. Abbreviated IPP. A company that generates power but is not affiliated with an electric utility.

induced current. Current in a conductor due to the application of a time-varying electromagnetic field.

induced voltage. A voltage produced around a closed path or circuit by a change in magnetic flux linking that path.

industrial sector. Electric utilities generally divide customers into classes, broadly, residential, commercial and industrial. The industrial sector includes manufacturing, construction, mining, agriculture and others. Industrial users generally have heavier electrical use than residential or commercial users.

insulator. A material that is a very poor conductor of electricity. The insulating material is usually ceramic or fiberglass when used in an electric line and is designed to support a conductor physically and to separate it electrically from other conductors and supporting material.

intermediate load. In an electric system, intermediate load refers to the range from base load to a point between base load and peak load. This particular stage may be the mid-point, a percent of the peak load, or the load over a specified time period.

internal combustion plant. A plant in which the prime mover is an internal combustion engine. This type of engine has one or more cylinders, in which the process of combustion takes place, converting energy released from the rapid burning of a fuel-air mixture into mechanical energy. Diesel or gasoline engines are the principal types used in electric plants. These plants are generally used only during periods of high electricity demand.

integrated resource planning. The process many utility commissions use to select the generation resources needed to meet future demand for electricity.

investor-owned utility. Abbreviated IOU. Electric utilities organized as tax-paying businesses and generally financed by the sale of securities. The properties are managed by shareholder-elected representatives. These are usually set up as publicly owned corporations. IOUs provide electric service to about 75% of the country's electricity consumers.

ionizing radiation. Radiation that comes from the high-energy end of the electromagnetic spectrum, such as x-rays, ultraviolet light, and gamma rays.

isotope. Chemical properties of elements are determined by the number of protons in the nucleus. Nuclear characteristics depend on the number of neutrons. Uranium (^{238}U) with 146 neutrons is stable while ^{235}U with 143 neutrons may be split with a slow neutron.

J

joint-use facility. A multiple purpose hydroelectric plant. An example is a dam that stores water for both flood control and power production.

Joule. A measurement of energy. It is the work done by a force of one Newton, when the point at which the force is applied is displacing one meter in the direction of the force. It is equal to 0.239 calories. In electrical theory, one joule equals one watt-second.

K

kilovolt. Abbreviated kV. Equal to 1,000 Volts.

kilowatt. Abbreviated kW. A measurement of electric power equal to one thousand watts. Electric power capacity of 1 kW is sufficient to light 10, 100-Watt light bulbs.

kilowatt hour. Abbreviated kWh. A measure for energy that is equal to the amount of work done by 1,000 watts for one hour. Consumers are charged for electricity in cents per kilowatt hour. One kilowatt hour is enough electricity to run 10, 100-Watt light bulbs for one hour.

L

lag. The delay between two events.

light oil. Lighter fuel oils distilled off during the refining process. Virtually all petroleum used in internal combustion and gasoline turbine engines is light oil.

light water reactor. The most widely used nuclear reactor type in the world. Ordinary water, called light water in the nuclear industry, is used as the coolant and moderator.

lignite. A brownish-black coal of low rank with high inherent moisture and volatile matter. Lignite is used almost exclusively for electric power generation. It is sometimes referred to as brown coal.

load. The amount of electric power required at a given time by energy consumers which can be divided into three major classes — industrial load, commercial load, and residential load.

M

megawatt Abbreviated MW. One million Watts.

megawatt-hour. Abbreviated MWh. One million Watts for one hour.

meltdown. The type of nuclear accident in which the fuel becomes so overheated that fuel in the fuel rods melts.

merchant plant. An electricity generating facility built and operated without long-term contracts guaranteeing sale of the electricity generated. Many such facilities are partial merchant plants, with contracts guaranteeing sale of a certain percentage of generation to a nearby utility.

municipal utility. An electric utility system owned and/or operated by a municipality that generates and/or purchases electricity at wholesale for distribution to retail customers generally within the boundaries of the municipality.

N

natural gas. A naturally occurring mixture of hydrocarbon and nonhydrocarbon gases found in porous geological formations beneath the earth's surface, often in association with petroleum. The principal constituent is methane (CH_4).

net capability. The maximum load-carrying ability of the equipment, exclusive of station use, under specified conditions for a given time interval, independent of the characteristics of the load. Capability is determined by design characteristics, physical conditions, prime mover, energy supply, and operating limitations, such as cooling and circulating water supply and temperature, headwater and tailwater elevations, and electrical use.

net generation. Gross generation less the electric energy consumed at the generating stations for station use.

neutral conductor. The grounded conductor in normal operations. Utilities and customers alike attempt to balance loads on hot conductors so current cancels out neutral conductors.

neutron. An uncharged particle found in the nucleus of every atom except hydrogen. Neutrons sustain the fission chain reaction in nuclear reactors.

nonattainment area. A geographic region in the U.S. designated by the Environmental Protection Agency as having ambient air concentrations of one or more criteria pollutants that exceed National Ambient Air Quality Standards.

nonutility generator. Abbreviated NUG. A facility that produces electric power and sells it to an electric utility, usually under long-term contract. NUGs also tend to sell thermal energy and electricity to a nearby industrial customer.

nonutility power producer. A corporation, person, agency, authority, or other legal entity that owns electric generating capacity and is not an electric utility. Nonutility power producers include qualifying small power producers and cogenerators without a designated franchised service territory.

North American Electric Reliability Council. Abbreviated NERC. Electric utilities formed NERC to coordinate, promote, and communicate about the reliability of their generation and transmission systems. NERC reviews the overall reliability of existing and planned generation systems, sets reliability standards, and gathers data on demand, availability and performance.

nuclear fuel. Uranium, plutonium, or thorium may be used. Light (ordinary) water-moderated reactors such as boiling water reactors (BWR) and pressurized water reactors (PWR) require enrichment of fuels to 3%-5% to achieve a self-sustaining reaction.

nuclear power plant. A facility in which heat produced in a reactor by the fissioning of nuclear fuel is used to drive a steam turbine.

nucleus. The positively charged core of an atom. All nuclei are made up of protons and neutrons, except hydrogen, which has only one proton.

O

off-peak power. Power supplied during designated periods of relatively low system demands.

Ohm. The unit of measurement of electrical resistance. Specifically, an Ohm is the resistance of a circuit in which a potential difference of one volt produces a current of 1 Ampere.

ohmmeter. An instrument for measuring electric resistance.

outage. The period during which a generating unit, transmission line, or other facility is out of service.

ozone. **(O^3)** A compound consisting of three oxygen atoms. It is the primary constituent of smog.

ozone transport. Ozone transport occurs when emissions from one area drift downwind and mix with local emissions contributing to the ozone concentrations in the downwind area.

P

peak days. In electricity, the days in the summer months when the demand for electricity is at its highest level due to air conditioning load. For natural gas, peak days are the days in the winter months when demand for gas is at its highest level due to most heating equipment being used.

peak-load plant. A plant, usually using old, low-efficiency steam units, gas turbines, diesels, or pumped storage hydroelectric equipment. These facilities are used only to fill in peak load during heaviest electrical use, such as the hottest days of summer, because they are more expensive to run than the baseload plants.

petroleum. A mixture of hydrocarbons existing in the liquid state found in natural underground reservoirs often associated with natural gas.

Potential hydrogen (pH). A measure of the acidity or alkalinity of a material, liquid, or solid. A pH of 1 is highly acidic; 7 is neutral; 14 is alkaline.

photovoltaic cell. A type of semiconductor device in which the absorption of light energy creates a separation of electrical charges.

power. (P) The instantaneous current being delivered at a given voltage, measured in Watts, or more usually kilowatts. Power delivered for a period of time is energy, measured in kilowatt-hours.

power pools. A group of utilities that coordinate the operation of their power plants and share the costs between themselves. Power pools are especially common in the northeastern U.S.

power marketer. A company that buys and resells power. These merchants typically do not own generating facilities.

power surge. A sudden change in an electrical system's voltage that is capable of damaging electrical equipment.

pressurized water reactor. Abbreviated PWR. A light water reactor in which the water used as a moderator is kept under pressure, preventing it from boiling at normal operating temperatures.

prime mover. The engine, turbine, water wheel, or similar machine that drives an electrical generator. Generally, a prime mover refers to a device that converts potential energy to kinetic energy, such as internal combustion engines, hydro or steam turbines, photovoltaic solar, and fuel cells.

private power producer. Any entity that engages in wholesale power generation or in self-generation.

privatization. The sale or transfer to private individuals or businesses of assets or businesses owned by the government.

privatization. Conversion of a government-owned firm or industry to private ownership.

proton. A particle in the nucleus of an atom with a single positive electric charge. The number of protons in the nucleus determines the chemical properties of an element.

public utility. Publicly owned electric utilities are nonprofit local government agencies established to serve their communities and nearby consumers at cost, returning excess funds to the consumer in the form of community contributions, economic and efficient facilities, and lower rates. Publicly owned electric utilities number approximately 2,000 in the U.S., and include municipals, public power districts, state authorities, irrigation districts, and others.

PUHCA. The Public Utility Holding Company Act of 1935. PUHCA regulates the large interstate holding companies that monopolized the electric utility industry in the early part of the twentieth century.

pumped storage hydroelectric plant. A plant that usually generated electric energy during peak-load periods by using water previously pumped into an elevated storage reservoir during off-peak periods when excess generating capacity is available. When additional capacity is needed, the water can be released from the reservoir through a conduit to turbine generators located in a power plant at a lower level.

PURPA. The Public Utility Regulatory Policies Act of 1978. PURPA promotes energy efficiency and increased use of alternative energy sources, encouraging companies to build cogeneration facilities and renewable energy projects. Facilities meeting PURPAs requirements are called qualifying facilities or QFs.

Q

qualifying facility. Abbreviated QF. A generator that 1) qualifies as a cogenerator or small power producer under PURPA, and 2) has obtained certification from FERC. They generally sell power to utilities at the utilities' avoided cost.

R

radiation. The process in which atoms and molecules undergo internal change, resulting in the emission of energy.

rate base. The value of property upon which a utility is permitted to earn a specified rate of return as established by a regulatory authority.

regulating transformer. A transformer used to vary the voltage, or the phase angle, or both, of an output circuit controlling the output within specified limits, and compensating for fluctuations of load and input voltage.

regulator. An electrical device that raises or lowers the voltage of the circuit to which it is attached.

regulatory compact. A theory advocated by some utilities which holds that in exchange for building the generation, transmission, and distribution infrastructure necessary to provide power to their service area, the utility is guaranteed a return on those investments.

regional transmission group. A voluntary organization of transmission owners, transmission users, and other entities approved by the Federal Energy Regulatory Commission to efficiently coordinate transmission planning and expansion, operation, and use on a regional basis.

regulation. The government function of controlling or directing economic entities through the process of rulemaking and adjudication.

REM. The standard unit of radiation dose. Radiation is often measured in millirems for low-level doses. Stands for roentgen equivalent man. See SIEVERT.

renewable energy. Refers to any source of energy that is constantly replenished through natural processes. Sunlight, moving water, geothermal springs, biomass, and wind are all examples of renewable energy resources used to generate electricity.

residential sector. The residential sector is defined as private household establishments that consume energy primarily for space heating, water heating, air conditioning, lighting, refrigeration, cooking, and clothes drying.

residual fuel oil. The topped crude of refinery operation. Residual fuel oil is used for the production of electric power and various industrial purposes.

rural electric cooperatives. Also known as co-ops, these entities are owned by their members. Co-ops were established to provide electricity to these rural members. The Rural Electrification Administration, part of the U.S. Department of Agriculture, was established in 1936 to extend electric service to small rural communities where it was ore expensive to provide electrical service. There are approximately 950 cooperatives in the U.S.

Rural Electrification Administration. Abbreviated REA. This agency was formed in 1936 to provide low-interest loans to expand electric service to rural areas.

S

scram. The rapid shut down of a nuclear reactor by moving control rods into the core to stop fission.

short circuit. An abnormal connection of relatively low impedence, whether made accidentally or intentionally, between two points of different potential in a circuit.

Sievert. (SU) SI unit for radiation dose. See REM.

small power producer. Under the Public Utility Regulatory Policies Act (PURPA) a small power production facility or small power producer generates electricity using waste, renewable, or geothermal energy as a primary energy source. Fossil fuels can be used, but renewable resources must provide at least 75% of the total energy input.

spent fuel. Nuclear fuel that can no longer sustain a chain reaction.

stranded benefits. Social programs and other regulatory "benefits" currently included in utility rates that could be stranded in an open market.

stranded costs/investment. Utility assets, mainly high-cost power plants, that would lose value in a competitive market.

steam electric plant. A plant in which the prime mover is a steam turbine. The steam used to drive the turbine is produced in a boiler where fossil fuels are burned.

stranded costs. This refers to a utility's fixed costs, usually related to investments in generation facilities that would no longer be paid by customers through their rates in the event that they opt to purchase power from other suppliers.

subbituminous coal. Also called black lignite, it is dull and black and usually contains around 20% to 30% moisture. The heat content of subbituminous coal ranges from 16 to 24 million Btu per ton as received. Subbituminous coal, mined in the Western coalfields, is used for generating electricity and for space heating.

substation. Facility equipment that switches, changes, or regulates electric voltage.

sulfur. One of the elements present in varying quantities in coal. Sulfur contributed to environmental degradation when coal is burned.

switchgear. A general term covering switching and interrupting devices and their combination with associated control, metering, protective and regulating devices, also assemblies of these devices with associated interconnection, accessories, enclosures and supporting structure used primarily in connection with the generation, transmission, distribution, and conversion of electric power.

switching station. Facility equipment used to tie together two or more electric circuits through switches. The switches are selectively arranged to permit a circuit to be disconnected, or to change the electric connection between the circuits.

T

transformer. An electrical device for changing the voltage of alternating current.

transmission. The process of transporting electric energy in bulk on high voltage lines from the generating facility to the local distribution company for delivery to retail customers.

transmission circuit. A conductor used to transport electricity from generating stations to load.

transmission company. A company engaged solely in the transmission function of the electric power industry.

transmission grid. An interconnected system of electric transmission lines that allows power to move from any point to another over multiple paths.

transmission line. A set of conductors, insulators, supporting structures, and associated equipment used to move large quantities of power at high voltage.

transmission system. An interconnected group of electric transmission lines and associated equipment for moving or transferring electric energy in bulk between points of supply and points at which it is transformed for delivery over the distribution lines to consumers or is delivered to other electric systems.

transmission grid. The high voltage wires that connect generation facilities with distribution facilities. It is the infrastructure through which power moves around the U.S. It is necessary to carefully coordinate use of the transmission system to ensure reliable and efficient service.

tritium. (H^3) A hydrogen nucleus with one proton and two neutrons. Tritium is used and produced in fusion reactions.

turbine. A machine for generating rotary mechanical power from the energy of a stream of fluid. Turbines convert the kinetic energy of fluids to mechanical energy through the principles of impulse and reaction, or a mixture of the two.

U

ultra-high voltage systems. Electric systems in which the operating voltage levels have a maximum root-mean-square ac voltage above 800,000 volts (800 kV).

unbundling. The process of separating natural gas services into components with each component priced separately. Traditionally, numerous gas services, such as sales, local transportation, and storage, had been tied together and offered to customers as a single, bundled product. By separating services into components, unbundling enables customers to compare the value of each service to its price. Unbundling also allows customers to choose those individual services that meet their own energy needs. This practice is expected to be part of the deregulation of the electric industry with generation, transmission, and distribution segments separated, as well as various value-added services and ancillary services.

uranium. The heaviest natural element enriched uranium is used as fuel for nuclear reactors.

usage rates. Segmenting consumers according to the volume of product they buy and the speed at which they use it.

utility. Privately owned companies and public agencies engaged in the generation, transmission, or distribution of electric power for public use.

V

value-added services. Services, such as security monitoring, telecommunications, internet access, and others, that add value to electric services. Other services can be offered by utilities to achieve greater customer satisfaction and loyalty.

variable costs. Those costs borne by electric utilities that vary with the level of electric output and include fuel expenses.

vertical disaggregation. Separating electric generation, transmission, and distribution functions of a utility into separate companies.

vertically integrated utility. Utilities that sell power on a bundled basis and whose activities run the full range of different functional activities of generation, transmission, and distribution. With deregulation of the electric utility industry well under way, vertically integrated electric utilities may well be on their way out.

Volt. The measure of electro-motive force (EMF) or pressure that pushes electric current through a circuit.

Volt-Ampere, reactive. Abbreviated VAR. A reactive load, typically inductive from electric motors, which causes more current to flow in the distribution network than is actually consumed by the load. This requires excess capability on the generation side and causes greater power losses in the distribution network.

voltmeter. An instrument that measures the electric potential difference between two points in a circuit in Volts.

W

waste-to-energy plants. A steam-turbine generating facility that uses municipal solid waste as the primary energy source to produce the steam used in the generating process.

Watt. The basic expression of electrical power or the rate of electrical work. One Watt is the power resulting from the dissipation of one Joule of energy in one second.

Watt-hour. An electrical energy unit of measure equal to 1 Watt of power supplied to, or taken from, an electric circuit steadily for one hour.

wheeling. The transportation of power to customers. Wholesale wheeling is transmitting bulk power over the grid to power companies. Retail wheeling is transmitting power to end users, such as homes, businesses, and factories.

wholesale wheeling. The use of transmission facilities of one system to transmit power by agreement of and for another system with a corresponding wheeling charge Wholesale wheeling involves only sales for resale and occurs when the buyer of the power resells the wheeled power to retail customers.

Bibliography

Asimov, Isaac, *The Intelligent Man's Guide to the Physical Sciences*, Pocket Books, Inc., New York, 1964.

August, Jim, PE, *Applied Reliability Centered Maintenance* (Tulsa, OK: PennWell Books, 1999)

Beaty, Wayne, *Electric Power Distribution Systems: A Nontechnical Guide*, PennWell Publishing Company, Tulsa, 1998.

Chambers, Ann, *Power Primer: A Nontechnical Guide From Generation to End Use* (Tulsa, OK: PennWell Books, 1999)

Cohen, Joel, E., *How Many People Can the Earth Support?*, W. W. Norton & Company, New York, 1995.

Del Toro, Vincent, *Electrical Engineering Fundamentals*, Prentice-Hall, Inc., Englewood Cliffs, 1972.

De Vries, Louis, *French-English Science Dictionary*, Third Edition, McGraw-Hill Book Company, Inc., New York, 1962.

Dickson, Paul, *The Official Rules*, Delacorte Press, New York, 1978.

Ebenhack, Ben W., *Nontechnical Guide to Energy Resources,* PennWell Publishing Co., Tulsa, 1995.

Famighetti, Robert, Ed. Dir., *The World Almanac and Book of Facts 1998*, K-III Reference Corp., Mahwah, 1997.

Freeman, Ira M., *All About Electricity*, Random House, New York, 1957.

Giampaolo, Tony, MSME, PE, *The Gas Turbine Handbook: Principles and Practices* (Liburn, GA: The Fairmont Press, Inc., 1997)

Graf, Rudolf F., Editor, *Dictionary of Electronics*, Fourth Edition, Radio Shack, Fort Worth, 1974.

Jackson, Herbert W., *Introduction to Electric Circuits*, Prentice-Hall, Inc., Englewood Cliffs, 1970.

Jordan, Edward C., Ed. in Ch., *Reference Data for Radio Engineers*, Seventh Edition, Howard W. Sams & Co., Indianapolis, 1985.

Kaku, Michio, *Visions*, Doubleday, New York, 1997.

Hutchinson, Charles L., Ed., *The ARRL Handbook for the Radio Amateur*, Sixty-Second Edition, American Radio Relay League, Newington, 1985.

Manes, Christopher, *Green Rage*, Little, Brown and Company, Boston, 1990.

Masterton, William L. and Slowinski, Emil J., *Chemical Principles*, Fourth Edition, W. B. Saunders Company, Philadelphia, 1977.

Moffett, George D., *Critical Masses: The Global Population Challenge*, Viking, New York, 1994.

Pansini, Tony, Smalling, K. D., *Guide to Electric Power Generation*, The Fairmont Press, Lilburn, 1994.

Payne, F. William, Editor, *Cogeneration Reference Guide* (Liburn, GA: The Fairmont Press, Inc., 1997)

Rosenberg, Paul: *Alternative Energy Handbook* (Liburn, GA: The Fairmont Press, Inc., 1992)

Seevers, O. C., P.E., *Management of Transmission & Distribution Systems*, PennWell Publishing Company, Tulsa, 1995.

Spiewek, Scott A., and Weiss, Larry, *Cogeneration and Small Power Production Manual* (Liburn, GA: The Fairmont Press, Inc., 1997)

Warkentin, Denise, *Electric Power Industry in Nontechnical Language*, PennWell, Tulsa, 1998.

Wildi, Theodore with McNeill, Perry R., *Electrical Power Technology*, John Wiley & Sons, New York, 1981.

1999 International Electric Power Encyclopedia (Tulsa, OK: PennWell Books, 1999)

Index

A

B

C

D

E

F

G

H

I

J

K

L

M

N

O

P

Q

R

S

T

U

V

W

Z